Financial analysis in agricultural project preparation

FAO INVESTMENT CENTRE TECHNICAL PAPER

8

by
K. Selvavinayagam
FAO Investment Centre

FOOD AND AGRICULTURE ORGANIZATION OF THE UNITED NATIONS
Rome, 1991

M-70
ISBN 92-5-103086-3

FINANCIAL ANALYSIS IN AGRICULTURAL PROJECT PREPARATION

TABLE OF CONTENTS

FINANCIAL ANALYSIS
IN
AGRICULTURAL PROJECT PREPARATION

PREFACE

Origin and Aims of the Paper

This paper grew out of a series of lectures given by the writer in April and September 1984 to FAO Investment Centre staff, mostly economists and project analysts.

Although a great deal has been written on the general topic of financial analysis, there is little literature on accounting for project planning and analysis. Where investment planning involves revenue-earning entities other than small farms, an analyst will need to know what set of forecast financial statements needs to be prepared for each entity for the life of the project, how these statements relate to each other and how they are put together in a consistent way in order to carry out different types of financial analyses. The paper therefore concentrates on those aspects of analysis which involve the use of financial statements.

The paper has been written principally to meet the needs of FAO Investment Centre economists and project analysts who have no previous training in accountancy, but may be of value also to persons of a similar background working on agricultural project preparation in developing countries.

Emphasis is placed on the discussion of concepts and techniques in order to provide users with an analytical framework for a systematic financial appraisal of a project. It is assumed that the user is familiar with the mathematics of discounting.

Scope of the Paper

The focus of the paper is on agricultural projects. The coverage is restricted to revenue-generating entities which are likely to maintain full sets of accounts, such as agro-industries, commercial farms and lending institutions. Farm budgets are thus excluded.

Organization and Structure of the Paper

The choice of an appropriate structure for the text has been difficult due to problems in achieving a right balance of theoretical concepts and practical applications. The actual choice is largely influenced by the writer's own experience.

The paper is structured into a main text, a glossary and a bibliography. Examples are provided in the text itself to illustrate the application of those financial principles and techniques discussed in the paper. The glossary is included to provide a broad definition of key terms that are referred to in the main text and that analysts may come across in practice.

Acknowledgements

The preparation of this paper has benefited from discussions with my colleagues in the Investment Centre particularly Edgardo Floto, Abdul Hameed, Pietros Kidane, Ricardo Uberti and several others who participated in the training courses. Andrew MacMillan and Duncan Knowler helped to review substantial parts of the final draft. I am also indebted to Rahul Raturi, presently with the World Bank and J.D. MacArthur, Senior Lecturer, Project Planning Centre, Bradford University, who have read an earlier draft and offered useful comments. All ideas presented in the paper have intellectual origins and lack of specific acknowledgement should not be taken to mean lack of awareness and gratitude for the contributions of earlier writers on the subject. I would like to express my thanks to the Chartered Institute of Management Accountants (London) for permission to reproduce parts of their official Terminology of Management Accounting. Hilda Richmond and Joan Price did a wonderful job in typing the draft with as much patience as my family had in tolerating me for turning leisure hours at home into working sessions.

FAO/World Bank Cooperative Programme
Rome
18 July 1991

K. Selvavinayagam

FINANCIAL ANALYSIS
IN
AGRICULTURAL PROJECT PREPARATION

I. INTRODUCTION

The Role of Financial Analysis

Financial analysis provides a practical means of assessing the profitability of investments and their likely financial impact on potential investors including farmers, lenders, profit-oriented enterprises like processing industries and marketing agencies, whether these are in the public or private sectors. It forms an integral part of project design because technical, economic, financial, commercial, social and institutional aspects are all interrelated. Although the requirements for and the basic principles of financial analysis do not vary between different sectors, there are some distinct problems in agriculture, which deserve special treatment. These relate mainly to: (i) short-term fluctuations in prices of agricultural commodities more than the prices of non-farm products; and (ii) the long-run tendency for farm incomes to fall below urban incomes. Farm crops, which are mostly seasonal, are subject to variability in output and prices (and hence incomes) because of many factors completely beyond farmers' control - weather, pests and diseases. In addition, primary commodities are more unstable than secondary ones, since they are less able to make the supply adjustments in the short run. The volume of production of tree crops like rubber, cocoa or tea is not much influenced in the short run by price. New planting can produce increased output only after several years, while once the tree is planted, it will go on yielding and it will be worth picking it so long as the price does not fall below the cost of collection and transport to the market. Accordingly if demand changes, there is little responsiveness on the supply side. These special problems of agriculture have implications for the stability of the farmer's income and his ability to invest. These problems are explored in the later chapters of this paper. For projects which do not generate revenue, such as education or research projects, financial analysis is concerned mainly with the sources and ascertaining the adequacy of funds for starting and maintaining the activity.

Purposes of Financial Analysis

Financial analysis is essentially undertaken for the following purposes:

(a) to determine the financial viability of a project or an enterprise;

(b) to assess the adequacy of a financing plan for a new project or a continuing business;

(c) to advise on methods of improving the viability of a project or enterprise including the appropriateness of tariffs, prices and cost recovery generally, and, on the financial arrangements, conditions or covenants which are required to support a loan; and

(d) to plan and control project/enterprise operations.

Financial viability. The concept of financial viability has two dimensions - the financial profitability and solvency of the existing operations of an enterprise, and the viability of a new project or enterprise. For a continuing enterprise, financial success would mean an ability to generate enough revenue to meet all financial obligations on a timely basis and command an adequate level of working capital for continued operations. Normally it also implies an ability to earn a reasonable rate of return on capital employed. The extent of success is ascertained by a review of its financial structure, liquidity trends, and profitability over time.

For a new project, the main objective of the analysis is to demonstrate that the financial cash flows expected to be generated by it are attractive to the prospective investors, inducing them to contribute equity funds to the particular project rather than to employ them elsewhere. The analyses on which such judgements can be based are the internal rate of return (IRR) and the net present value (NPV). An investor would be interested in the IRR after tax which he would compare with returns from alternative investment opportunities at similar risk levels before committing funds to a particular project. If the IRR after tax of the project is greater than the cost of capital, it can be concluded that the project is viable. It must be emphasised that this will only be so if the machinery for the recovery of funds is appropriate and the terms of repayment can be adjusted to take any problems of phasing into account. IRR does not provide any information on the requirements for phasing, short-term bridging finance or grace periods on the loan required to accommodate delayed benefits. This leads us to the next objective of financial analysis.

Financing plan. The investment expenditures, including the incremental working capital needs of a project/enterprise, will have to be met on a timely basis and at a minimum cost. In drawing up the financing plan, consideration will have to be given to the various types of capital and to the most effective structure, both of which would be coordinated to satisfy the financial requirements of the business. These requirements are to be ascertained by careful budgeting to avoid extremes of over-capitalization or under-capitalization. In a continuing business where a budgetary control system is in operation, the forecasting of requirements should present no difficulty. For new projects, more has to be left to conjecture. Not only must costs be estimated but revenue is dependent on demand, which, in turn, is influenced by the state of the domestic economy, both of which lie largely outside the control of the enterprise. In both new and existing businesses, funds may be raised from external sources but in a continuing business, internal resources can be mobilized by ploughing back profits.

Certain factors need to be considered when external financing is contemplated. These are: **the trend and seasonality of a firm's funding requirements, the financial condition and performance of the firm** and **the inherent risk of business operations.** In agriculture, risks arise from weather, fluctuating prices, and plant or animal diseases. These are reflected in variations in income flow (some companies are in stable lines, e.g. an electricity board, while others are in highly volatile lines, e.g. a rubber estate). All three of these factors should be considered in determining the financing needs of the enterprise. Moreover, they should be considered jointly. The nature of the need for funds influences the type of financing that should be used. If there is a seasonal component to the business, it lends itself to short-term financing and bank loans in particular. The financial condition and performance of the firm also influence the type of financing that should be used. The greater the liquidity, the stronger the overall financial condition and the greater the profitability of the firm, and the more risk that can be incurred with respect to pattern of financing. In such situations debt financing becomes more attractive. The basic business risks faced by the firm also have an important bearing on the type of financing that should be used. The greater the business risk, the less desirable debt financing usually becomes relative to equity financing. Equity financing is safer in that there is no contractual obligation to pay interest and repay principal as there is with a loan.

The figure below presents the interrelationship between the above factors. The last box in the figure recognizes that it is not simply an exercise in determining the optimal financing plan from the standpoint of the firm and assuming that it can be fulfilled. The plan needs to be sold to external suppliers of capital. In the end, the firm may have to modify its plan to meet the realities of the market place. Interaction of the firm with the suppliers of capital determines the amount, terms and cost of financing.

Financing Requirements of the Firm

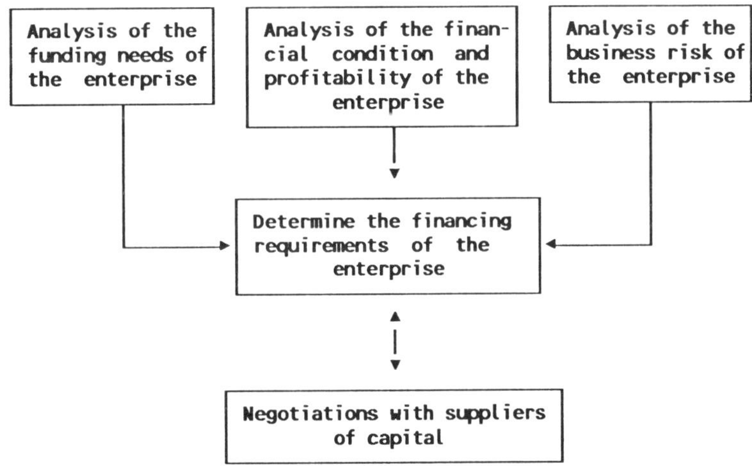

Financial planning and control. In order to generate both increasing profits and a satisfactory return on assets employed, firms must plan their growth. To minimize the cost of financing the planned growth, they must also plan the timing, source and volume of additional financing to be raised. Thus, financial planning and control should be an integral part of a firm's profit planning and control system. In such a system, planning is the basis of control and the control process is broadly one of:

- analysing historical performance;
- assessing the future environment in which the firm will be operating;
- developing long-term objectives, including financial objectives;
- formulating a strategy to achieve the objectives;
- translating the strategy into operating plans for the next 3 to 5 years and more detailed budgets for the next year;
- motivating employees to achieve the plans and budgets;
- periodically comparing actual with planned performance and reporting to responsible management as a basis for improving managerial efficiency and the effectiveness of the planning process.

As illustrated below, this is a continuous process. Within this cycle, firms must seek out and appraise investment opportunities (long-term) and make various tactical (short-term) decisions on such issues as pricing and product mix.

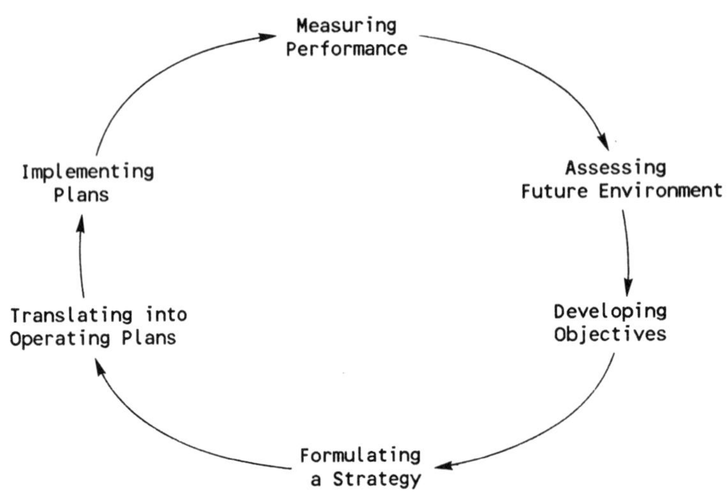

Several management accounting techniques are used to facilitate planning and control of business operations. The most popular ones are budgetary control, standard costing and network analysis.

Financial control systems may be divided into:

(a) financial control of projects;

(b) financial control of departments in an enterprise or simply responsibility centres.

There will be a continuing interest in the control of responsibility centres as departments, like their parent enterprise, are assumed to be in operational existence for the foreseeable future; in contrast, control over a project is limited to its life, a defined period of time. Also, information requirements of projects and responsibility centres differ. Responsibility centres can be expense centres (e.g. extension unit in a service cooperative) where the allocated financial responsibility of the manager is confined to costs, or profit centres (e.g. bank branch) where the manager of the centre is responsible for profit, or investment centres (e.g. livestock or crops in a mixed farm) where the manager is given responsibility for both profit and investment.

Project control implies that the selected projects are effectively implemented. During the course of implementation, the management are likely to be interested in four related questions.

- Is the project 'on time' and will it be completed as planned?
- Will the result of the project be as previously projected?
- Is the cost going to be as estimated?
- Is there a need for an adjustment in plans?

Financial performance. Should a weak financial position emerge during the course of appraisal of a project or its parent enterprise, a thorough investigation of the causal factors must be undertaken. Normally, this will be a multi-disciplinary exercise because the reasons for poor performance may go beyond financial factors. Taking financial factors into account, it can be said that pricing (including cost recovery) policies, cost reduction and cost control efforts exert a significant influence on financial performance, as do financial arrangements, particularly financial covenants. Financial covenants provide a framework for financial discipline and act as yardsticks in promoting better performance of revenue-generating enterprises. In cases where attainment of required financial standards is not practicable in the immediate future, the covenants serve as a statement of objectives for raising the level of performance to the desired level over a reasonable time-frame.

Financial covenants may be classified in terms of the revenue, capital structure and liquidity of a project/enterprise. Revenue covenants include those covering the rate of return, operating ratio, break-even point and cash generation; capital structure covenants are related to the debt-equity ratio, debt service coverage ratio, absolute debt limitation, and dividend limitation; liquidity covenants consist of the current ratio and quick asset ratio.

Relationship between Financial and Economic Analysis

There are two ways of assessing the desirability of undertaking a project - financial and economic analysis. The underlying tools used for carrying out financial and economic analysis are not different and both types of analyses are required for project screening and selection. However, there is a difference in approach since financial analysis deals with the cost and benefit flows from the point of view of a firm or individual as opposed to economic analysis which deals with the costs and benefits to society. Economic analysis, therefore, takes a broader view of costs and benefits than financial analysis. The methods nevertheless differ in several important ways. An enterprise is interested in financial profit and the stability of that profit, while society or government is concerned with much wider objectives such as food self-sufficiency, rural employment, poverty alleviation, and resulting net benefits to society as a whole. Therefore, the objectives of the two types of analysis are different.

The two analyses also differ on account of the basis used for valuing inputs and outputs from a given project. The resulting costs and benefits are not necessarily the same under the two types of analysis. Financial analysis includes as costs all payments that reduce the monetary resources of the project, and considers as benefits (or revenues) all receipts that increase the project's financial resources. Economic analysis treats as costs only those payments which reduce the nation's real resources, and as benefits only those receipts which increase the nation's real resources. Monetary resources are distinguished from real resources by the fact that there are certain payments (e.g. taxation) and receipts (e.g. unemployment benefits) which are in the nature of 'transfers' from one section to another section of the society, and do not in any way affect the total availability of real resources to the economy as a whole. **Taxes** and various forms of **subsidies** are examples of such transfer payments and receipts. However, these payments and receipts form an integral part of financial analysis since they change the availability of monetary resources to the projects under consideration. Thus, a company tax will decrease profits while an investment grant will augment the monetary resources of the enterprise.

The financial and economic costs may also differ considerably, for example, in the values attached to imports into a country with an overvalued exchange rate; to the value placed upon labour in conditions of underemployment; or to intangibles such as pollution which may have no financial costs to the enterprise causing it but a high cost to society.

The distinction between financial and economic pricing may be illustrated by the following examples. Consider the price of petrol at $5 per gallon which contains a large element of value added tax (25%). The price to be used for financial analysis will be $5 per gallon, while for economic analysis it will be $4, i.e. net of tax $1. Similarly, subsidy elements in prices should be adjusted to reflect economic costs. For example, consider a market price for fertilizer of $50 per ton used by a coconut estate. Given a subsidy of 50%, the financial cost to the estate will be $50; the economic cost will be $100. A third illustration is in respect of labour input. If the wages which have to be paid to the unskilled labour on an investment project are $3 per man-day, which is greater than their marginal opportunity cost, then financial analysis will use $3 per day, while economic analysis will discount it by a margin to bring it to its opportunity cost.

The difference in pricing is illustrated as follows:

	Financial Analysis	Economic Analysis
I. For Subsidies	Subsidized price or cost	Exclude Subsidies
II. For Taxes	Tax-inclusive price or cost	Exclude Taxes
III. Opportunity Cost		
- Inputs/labour	Local farm-gate price	Opportunity costs
- Capital	Cost of borrowing or return which could be earned by investing capital elsewhere than in the project	Opportunity cost of capital
- Project-generated output	Local farm-gate price	World market parity price/border price

Alternatively, it is possible to make the distinction in a slightly different manner by drawing attention to the occasions where revenues and costs of an enterprise are a poor representation of social gains and losses. There are four groups of circumstances which can make business revenues and costs a poor measure of gains and losses from a national viewpoint. The first is that the output of services provided may not be marketed, as for instance, in coastal protection works. Secondly, some of the gains or losses may accrue to third parties, as in salmon fisheries, where a change in the type or extent of fishing at river mouths affects rod fishing upstream. Thirdly, as indicated earlier, the market value of inputs is, for instance, affected by subsidies. Finally, a proposed investment scheme could make such a large impact upon a market that the ruling market price is affected by it. A decision then has to be made whether the old or new price or some combination, provides the relevant valuation for economic evaluation. Several adjustments may thus have to be made to convert the financial costs and benefits into social costs and benefits. In this sense, data generated for financial analysis can serve as a primary source for economic evaluation.

As in economic analysis, financial analysis is not a one shot affair at preparation or appraisal of a project. Implementation of plans raises new issues. This is especially so in projects with flexible plans. Financial analysis should then be an important feature in adapting plans and redesigning projects. Moreover, financial results should contribute to improved project design by demonstrating the incentives or disincentives which will exist to spur or discourage investors to participate in the project. Thus, for instance, a soil conservation practice may be shown to be of limited financial attractiveness to farmers but of considerable economic interest from a national point of view: such analysis provides the rationale for introducing funding arrangements to bridge the gap between individual and national interests.

II. FRAMEWORK OF FINANCIAL ANALYSIS

General Principles of Financial Analysis

Whatever the type of investment, it requires the outlay of cash during an initial period which is normally followed by periods of cash inflows from revenues and cash outflows to meet operating costs; some of the investment costs may also be incurred over the future. There is thus a distinct 'time-profile' of benefits and costs for each of the investment projects under consideration. The time-profile for one project may have large net inflows during, say, the first few years, and small net inflows thereafter. For another project, the time-profile may be the reverse of this. Yet a third project may have more modest net cash inflows spread evenly over a longer period than the other two.

The formulation which, as a description, best covers most financial analysis is as follows: the aim is to maximize the present value of future net cash flows. This formulation is very general but it does at least enable the analyst to set out a series of questions, the answers to which constitute the general principles of financial analysis. The financial analysis requires a proper identification of all cash cost and revenue streams over the project life, drawing a distinction between: (a) **capital cost stream**; (b) **operating cost streams**; and (c) **revenue streams**. The difference between the costs and revenues for each year results in the net benefit or **net cash flow** for each year. The stream of annual net benefit (or net cash flow) is generally referred to as the net benefit stream. The main questions for this analysis include:

(1) **What are the costs to be included?**
(2) **What are the revenues to be included?**
(3) **At what interest rate (i.e. cost of capital) are the costs and revenues to be discounted?**
(4) **What is the estimated life of the proposed investment?**
(5) **What is the logical framework for analysis?**

Three General Issues

Before taking up these questions, it is convenient to discuss three issues which involve more than one of these questions - the **calculation of incremental cash flows**, the **methodology of financial projections** and the **problem of risk.**

Incremental cash flows. Where proposed investment is for a new entity, the net cash flow stream of the project will all be incremental. However, where investment leads to an addition in the operations of an existing entity, incremental flows will have to be derived by comparing the net cash flows 'without' the project (i.e. assessing the future of only the existing facilities without the proposed investment), and net cash flows 'with' the project (i.e. assessing the future of the facility including new investment). A comparison, on the basis of 'before-and-after' approach should be avoided. The 'before-and-after' approach focuses attention on two points in time which may be misleading.

The life of most investment projects usually exceeds 10 years and it may not be prudent to assume that resource productivity will remain completely static during this time. Therefore to take the 'before' level of output from a set of resources may understate their true opportunity cost.

Financial projections. Financial projections are required to quantify cash inflows and outflows of participating entities to demonstrate their viability and credit-worthiness. Financial projections also help to identify external funding needs of these entities. For many agricultural projects, typical farm budgets can be prepared to determine the farmers' incentive to participate and their capacity to service loan-financed investments. For revenue-earning projects with marketable outputs, such as an industry or an agricultural plantation, annual financial projections for the project should be prepared showing revenues and costs of operation, including interest on borrowed funds. For a project which is a part of a large investment package rather than a self-contained investment, a separate treatment of project expenditures and revenues may not be practical and the analysis should focus on the soundness of the overall operations of the implementing agency. For projects implemented by autonomous or semi-autonomous entities, a comprehensive analysis of the entity's past and forecast financial performance must be undertaken. This analysis should cover all activities of the entity, including its existing operations, the proposed project, and any other planned investments. Annual financial statements projecting these activities should include income statements, cash flow or funds flow statements and balance sheets.

Financial projections will have to be critically reviewed as to their scope and coverage, underlying assumptions, length of the period covered and the reasonableness of their forecasts. The projections would normally include a full set of financial statements supplemented by forecasts of production, sales, operating expenses, trade debtors, creditors and interest on borrowings. Various methods of projection are described in Appendix 1. For calculation of the return on investment, the projections would be expressed in constant prices, i.e. as of a certain date, and for estimation of the external financing gap, current prices will be used. The projections should cover at least three to five financial years from the date of the proposed investment, and the underlying assumptions should be carefully reviewed to ensure that the forecasts are reasonable.

Risk. There are various ways in which risk impinges on financial analysis. The first point is that investment projects are not free of risk. The second is that allowances for risk can be made: (i) in the assessment of annual levels of costs and revenues; (ii) in the assumptions about length of project life; and (iii) in the discount rate. The first is appropriate when the risk of dispersion of outcomes is irregularly distributed with time. If the main risk is that revenues may at some point disappear or costs soar, the second type of adjustment would seem correct. The third correction, a premium on the discount rate, is appropriate where risk is strictly a compounding function of time.

Enumeration of Costs

Quantitative estimates of cost should be prepared annually for each project over its economic life. These estimates should distinguish between **capital costs** for: (a) fixed assets including land, civil works, buildings, equipment and machinery, but also engineering services and in-plant infrastructure; (b) outside plant infrastructure; (c) start-up costs; (d) working capital; and (e) replacement of fixed assets, when this is required within the life of the project.

For self-constructed fixed assets, the financing costs involved with the construction before the asset is completed may become material and a strong case for their inclusion in the acquisition costs can be made, as they are part of getting the asset ready for use. It can also be argued that borrowing costs incurred as a consequence of a decision to acquire an asset are not intrinsically different from other costs that are commonly capitalized. If an asset requires a period of time to bring it to the condition and location necessary for its intended use, the borrowing costs incurred during that period as a result of expenditure on the asset are a part of the cost of acquiring the assets. **Start-up costs** are one-time costs that are specifically concerned with the project. Examples of these costs are legal fees, payment for feasibility studies, architect's or surveyor's fees.

The working capital represents the excess of current assets over current liabilities. Most investment projects will necessitate outlays on current assets (e.g. inventories and debtors) and current liabilities in order to support the higher level of operations resulting from the new investment. For some projects, the working capital required will exceed the initial fixed investment (e.g. companies leasing agricultural and other equipment). As the level of activity eventually falls, so does the working capital requirement until, at the end of the project's life, the remaining working capital is recovered (see Chapter X).

Operating Cost Estimates

Operating cost estimates (also known as current or revenue expenditure) are based on input requirements and capacity utilization. These costs can be classified into variable and fixed costs and within each category they can be further divided into production costs, administration costs, marketing costs and research and development costs. This distinction permits estimates to be made of the operating costs at different production levels. Typically, operating costs are estimated for the 'full development' stage in the production level and pro-rated for other levels having regard to the fixed nature of some of the costs. Examples of such operating costs are cost of sales, wages, salaries, rent, rates, electricity etc. Some revenue expenditure may be carried forward (unsold stock) and charged in the following periods when the sale is made in accordance with the accruals concept which governs matching costs against revenue. Other items of revenue expenditure (research and development) are sometimes carried out forward on the grounds that the corresponding benefits will accrue over a number of subsequent accounting periods. Alternatively, they may be treated as capital costs.

Revenue Estimates

Revenues may arise from the sales of goods or services. Sales revenue should be the cash value of sales expected to be realized during the accounting periods under consideration. If the sale is on credit, adjustment is made to the Accounts Receivable shown under incremental working capital. The physical volume and price per unit underlying these calculations should also be shown.

In certain cases, the effect of a new investment may extend beyond the confines of that investment's cash flows. For example, a new investment by a fruit processing company to expand capacity and sales for lychees may result in a reduction in the sales and costs of the competing product - rambuttan - also processed by the company. Such effects should be included in the analysis.

Validity of cost and revenue estimates. The estimates upon which investment decisions are based are subject to errors arising from misjudgement or changed circumstances. The main sources of errors are: (i) the project may cost more or less than expected; (ii) working capital assumptions may not be achieved (e.g. credit terms or stock levels).

Errors in trading estimates are that: (i) business activity (total market) may differ from expectations; (ii) the undertaking's share of the market may not be realized; (iii) price level assumptions may be incorrect; (iv) costs of operation may differ from the forecasts.

Unforeseen non-commercial factors include: (i) government control or interference, which may also affect some or all the commercial factors; and (ii) alterations in the tax structure.

The extent of possible error will, of course, vary with the type of business, but an analyst familiar with a particular industry ought to be able to fix roughly the limits within which variations could occur.

Choice of Interest Rate

Revenues are normally generated only in the future while the costs are incurred immediately as the project investment is made. The investor has therefore to compare the value of future income receipts with current costs. If the enterprise borrows now to finance the investment, the value of future benefits must be sufficiently large relative to the current costs to pay the interest on the loan financing the investment. If investors' own funds were used, the interest rate will still matter because he should make the best use of the money on hand. It could have earned a higher return if it were used to purchase bonds or deposited with financial institutions. Thus, the interest rate relevant to investment decisions is the **real** interest cost of borrowing investment funds or return which could be earned by investing capital elsewhere.

The Estimated Life of the Investment

Estimation of length of life is clearly a highly subjective process depending on assessments of the physical length of life, technological changes, shifts in demand, emergence of competing products and so on. The criterion is the continued ability to generate satisfactory cash flows. In principle, there is an optimal life for an investment asset related to the relative costs of repair and replacement and closely connected to technological development. For example, new technology may render an investment uneconomic well before the asset's economic life. Moreover, the discount factors decline over the years so that the effect of cash flows in later years, say after 20 years, on project's IRR - becomes less material.

Analytical Framework

Full set of accounts. For the purpose of an enterprise's financial analysis, it is important to use a consistent framework for developing financial projections and determining cost and revenue streams. The analytical framework normally used in the above process is a set of financial accounts Financial statements are therefore the starting point of enterprise financial appraisal. The term 'financial statements' refers to balance sheets, income statements or profit and loss accounts, statements of changes in financial position (which may be presented in a variety of ways, for example, as a statement of cash flows or statement of source and application of funds), notes and other statements and explanatory material which are identified as being part of the financial statements. These statements are projected for the life of the project to derive the cost and benefit streams of the project. In the case of a new entity, the preparation and analysis of the financial statements constitute the full task. In the case of existing enterprises, the financial position will need to be assessed from analysis of their declared accounts, whether audited or not.

Quality of financial statements. Before starting an analysis of the financial statements, their usefulness and reliability must be checked by their consistency with: generally accepted accounting principles (as defined by the International Accounting Standards Committee [IASC]); national accounting policies; the legal and regulatory framework; inflation accounting (for countries experiencing inflation rates exceeding 30%) based on Accounting Standard 29 of the IASC; an auditor's report, particularly when qualified, for example, in respect of a change in accounting policies, inadequate provisions for losses and unrealistic revaluation of assets; and the reasons for change of auditor, if any.

Reliability of financial statements. To ensure reliability of these statements, it is necessary to have them audited by auditors who are independent (of the control of the entity to be audited and of the person appointing them), experienced, competent and reputable, using procedures and methods that conform with the relevant national standards or practices established within the country on the entity's annual financial statements which are prepared in accordance with International Accounting Standards or relevant national standards for businesses. This process of examination and verification (i.e. audit) normally results in a written opinion and report by the auditor, indicating the extent to which the financial statements and supporting information reports provide a true and fair view of the financial condition and the financial performance of the enterprise. Audited statements for at least three to five financial years and the accompanying notes should be detailed enough to allow meaningful analysis. Provisional financial statements can be accepted pending audit provided they are certified by the management.

III. METHODS OF INVESTMENT APPRAISAL

Criteria for Investment Decision

For a proper comparison of the value of different projects, it is clearly necessary to reduce all their cost and benefit streams to a single figure. A common method of appraisal is therefore required, which can be applied to a range of investment decisions and which should assist in deciding whether any particular investment will help the firm in maximizing the present value of future cash flows. In practice, most firms use one or more of the following methods.

(a) **Payback;**

(b) **Accounting rate of return (ARR);**

(c) **Net present value (NPV)**; and

(d) **Internal rate of return (IRR).**

Before examining these methods, two points must be noted. First, it must be emphasized that none of these methods will make a decision for the manager to invest or not invest, but can only serve to provide information to the decision-maker. A final decision will be based on a whole range of considerations beyond the confines of these formulae, extending to managerial judgement as to which of these methods contribute the soundest information base. Secondly, the key to good investment decision-making lies in the quality of information underlying the investment proposal. The actual method of evaluation is of only secondary importance compared with the critical issue of the reliability of the supporting estimates and assumptions. **It is of little benefit to use sophisticated appraisal techniques unless the degree of accuracy they suggest is matched by the quality of data used in the analysis.**

To illustrate the various methods, one example will be used throughout; the following data apply:

(1) A pineapple processing company proposes an investment involving an immediate cash outlay on a machine costing $30,000.

(2) Net cash flow (before depreciation at 20% p.a. straight line), is:

Yr 1 - $5,000
Yr 2 - $9,000
Yr 3 - $12,000
Yr 4 - $4,000
Yr 5 - $10,000 including salvage value of machine at $6,000.

(3) The machine will last five years which can be assumed to be the economic life of the project; the salvage value being $6,000.

The Payback Method

The payback method is also known by other names including **payback period, pay-off period, capital recovery period** and **cash recovery factor**. It is defined as the number of years required for the stream of cash flows generated by an investment to equal the original cost of that investment. It can be used to assess project acceptability as follows: (a) in accepting a project if the payback of an initial investment outlay is obtained within a predetermined time; and (b) when a comparison is required of the relative desirability of several mutually exclusive investments. In the latter case, projects can be ranked according to speed of payback with the fastest paying back project being the most favoured.

Using the above example and assuming the criterion of project selection as four-year (maximum) payback, the pineapple company investment is considered acceptable because it pays back the initial outlay of $30,000 within the stipulated time period, as shown below.

Year	Net Cash Flow before Tax ($)	Cumulative Cash Flow ($)
0	(30,000)	-30,000
1	5,000	-25,000
2	9,000	-16,000
3	12,000	-4,000
4	4,000	payback 0
5	10,000	10,000

The payback period is quick and simple to compute and is readily understood by management. The speed with which investment costs are recovered, as conveyed by this method, could become a major consideration for companies with limited availability of funds. Furthermore, the emphasis on speed of return falls in line with management's wish to avoid or minimize risk.

Five fundamental problems arise with this method.

(i) It ignores cash flows occurring after the payback period, as cash flows of $10,000 (profit $4,000 plus salvage value $6,000) arising in Year 5 are not taken into account. This exposes the analysis to some doubt, in ranking projects with varying cash flows within and beyond payback period.

(ii) It fails to allow for the timing of cash flows by treating cash flows in all years as equally important. However, this difficulty can easily be overcome by applying the method, not to the actual cash flows, but to the 'present value' (or discounted) cash flows. This technique is known as the 'discounted payback' method. All the same, there is another problem with the method in that there are no specific guidelines on how the maximum payback period should be determined.

(iii) The cash flow used in this method is the figure of annual profit before tax and depreciation. Normally, profit after tax but before charging depreciation, is treated as the relevant cash flow for this method.

(iv) The method favours projects having higher cash flows in early years. However, in many projects cash flows are slow to start for a number of reasons, which include gradual build-up of production, impact of marketing effort and phasing of development expenditure.

(v) No formal assessment is given of risk. Estimates of cash flows are likely to be less reliable further into the future. The payback method assumes that all cash flows are equally certain.

Despite these disadvantages, it remains one of the more popular appraisal methods. Nevertheless, it cannot really be considered as a particularly sound investment appraisal technique because of its major defect - its inability to report the expected return over the whole life of an investment, through its disregard of the investment's post-payback period flows. However, the payback method does provide a useful 'rule of thumb' check for the various minor investment decisions that companies have to make, which are financially too small to justify effort in using more complex but theoretically more correct appraisal methods. At best, for large investments, it can be useful only as an initial screening device before more appropriate methods are applied.

Accounting Rate of Return

The accounting rate of return (ARR), like payback, is another traditional measure of investment appraisal; it has several names and equally different methods of computation. A most common name is **return on capital employed**. This is calculated in a number of ways. In its basic form, it is calculated as the ratio of the accounting profit generated by an investment project to the required capital outlay, expressed as a percentage. The ARR is normally expressed as the ratio of the average annual profit generated over the life of the investment to the average annual capital employed, and stated as a percentage.

Using the pineapple company example, average capital employed and average annual profit are calculated.

Average capital employed[1] = $\frac{\$30,000 + \$6,000}{2}$ = \$18,000

Average annual profit[2] = $\frac{\$(200+4,200+7,200-800-800)}{5}$ = \$2,000

ARR = $\frac{2,000}{18,000}$ x 100 = 11%

---

1 Value of the investment in the machine at the beginning (\$30,000) and its scrap value (\$6,000) at the end.

2 Computed after deducting depreciation of \$4,800 each year (30,000-6,000 ÷ 5).

The ARR calculation serves as an investment decision guide in two ways. First, it provides a straightforward 'accept or reject' decision by providing a rule that projects may only be accepted if their ARR exceeds some 'target' rate of return. Second, where projects have to be ranked in order of preference, indication of a higher ARR by a project would suggest a greater preference for it to be accepted.

The method is superior to the payback period because of its advantage of not only being relatively quick and easy to calculate, but also of producing a percentage rate of return, most familiar to financial management of the company. It also overcomes the important defect of the payback method by taking account of the profit flows over the whole of a project's life. Moreover, management's success or failure in taking financial decisions in aggregate is judged among other things on the basis of the company's ARR. Therefore it would appear sensible that individual investment decisions should be taken on the basis of that same appraisal technique. Nevertheless, there are a number of disadvantages.

(i) The most important defect is that the ARR fails to recognize the time value of money and concentrates on a project's financial flows in accounting terms rather than cash flows.

(ii) The concept of ARR is ambiguous. There exist many variations, with little general agreement, on how capital employed should be calculated or on how profit should be defined. Some variations are discussed in Chapter XII.

(iii) The method measures the worth of a proposed investment in percentage terms. It is therefore unable to take into account the financial size of a project when alternatives are compared.

(iv) Although the technique allows for profit flows over the life of a project, it ignores the possibility of differing length of lives. When comparing alternative projects, a difference in length of life may be of critical importance for a number of reasons, e.g. uncertainty, liquidity.

However, in spite of these criticisms the method is widely used. Like payback, it is still capable of providing an acceptable basis for deciding on relatively minor short-run investment projects.

The above review shows that, apart from the possible exception of the appraisal of relatively small, shortlived investments, neither method has sufficient advantages to offset its disadvantages, and particularly the failure of both to allow for the time value of money is a significant weakness.

Time Value of Money

Unlike the two traditional methods of appraisal (payback and ARR), two other approaches in use - net present value and internal rate of return - explicitly allow for the time value of money through a process of compounding/discounting cash flows. In order to fully understand these methods, it is necessary to clarify the concept of the time value of money. The principle is that $100 received today is not equal to $100 received one year later. This is not due to a decline in the value of money in real terms, i.e. inflation, but simply to the existence of alternative investment opportunities. Money received today can be invested to earn interest over the ensuing year. If the interest rate is 10% p.a., $100 invested now will give $110 one year later. In other words, $100 today is equal to $110 one year from today at an interest rate of 10% p.a. Alternatively, $100 one year from today is equal to $90.91 today, its present value, because $90.91 plus interest at 10% for one year amounts to $100.

The mathematics of this process are based on the compound interest formula:

$$A = P(1 + r)^n$$

where **A** is the future value (or the sum of initial amount **P** plus interest), **P** is the initial amount, **r** is the annual rate of interest and **n** is the number of years for which **P** is invested.

$$A = \$100(1 + 0.10)^1 = 100 + 10 = 110$$

Restating the above formula, it can be used to calculate the present value as follows:

$$P = \frac{A}{(1 + r)^n}$$

$$P = \$110 \times \frac{1}{(1 + 0.10)^1} = \frac{110}{1.10} = \$100$$

Therefore $100 is the present value of $110 received in one year's time assuming annual interest at 10%.

Similarly, \$100 invested for 2 years at 10% p.a. would yield:

$$A = \$100(1 + 0.10)(1 + 0.10) = 100(1 + 0.10)^2 = \$121$$

and \$121 received in two years' time is equal to:

$$P = \$121 \times \frac{1}{(1 + 0.10)(1 + 0.10)} = \frac{121}{(1 + 0.10)^2} = \$100 \text{ now}$$

Given the time value of money, it is clear that money at different points in time is not directly comparable. For a proper comparison of the value of money at different points in time, it is clearly necessary to reduce all these time-profiles to a single figure. To do this, two main measures using the discount cash flow method are used. They are the net present value and the internal rate of return.

Net Present Value

The NPV method consists of discounting all future cash flows to the present value by means of some appropriate rate of interest. The rate of interest to be used should reflect the minimum rate of return which is acceptable to the firm, for a given investment. It works on the simple but fundamental principle that an investment is worth undertaking only if the present value of the cash inflows is at least equal to, if not greater than, the present value of the cash outflows arising from an investment. To put it another way, companies should make investments in projects with a zero or positive net present value.

The quantitative data required to carry out this method are: the initial cost of the project; the cost of supplying the capital required or the minimum rate of return acceptable; the value of the future cash flow in each period; the life of the project; and the discount factors.

The computation is carried out as follows:

(i) Calculate the present value of each year's net cash flow by multiplying the projected cash flow by the appropriate discount factor.

(ii) Add the computations to arrive at the single figure of net cash flows in present value.

If the result of (ii) (which is NPV) is zero or positive, i.e. the NPV of the cash flows is at least equal to, if not greater than, the NPV of all cash outflows, the investment would be acceptable at the discount rate used; if the NPV is negative, the investment should be rejected.

Considering the pineapple processing company example, and assuming a 10% as the minimum acceptable rate of return, calculations below show that the investment produces a NPV of \$350, suggesting that the proposal can be accepted.

Year	Cash Flow after Tax	DCF Factors (Discount Factor)	Present Value Cash Flow
	($)		($)
0	(30,000)	$(1 + 0.10)^{0} = 1$	(30,000)
1	5,000	$(1 + 0.10)^{-1} = 0.9091$	4,955
2	9,000	$(1 + 0.10)^{-2} = 0.8264$	7,438
3	12,000	$(1 + 0.10)^{-3} = 0.7513$	9,016
4	4,000	$(1 + 0.10)^{-4} = 0.6830$	2,732
5	10,000	$(1 + 0.10)^{-5} = 0.6209$	6,209
		NPV =	**$350**

The NPV method can be represented by the mathematical formula:

$$NPV = \sum_{n=1}^{n} CF_n(1 + r)^{-n} - C$$

Where $\mathbf{CF_n}$ is the cash flow in any period **n** (e.g. Year 1,2,3..Year **n**), **r** the rate of return or cost of capital and **C** the capital cost of the project. The table of discount factors provides the value of the expression $(1 + r)^{-n}$ for required values of **r** and **n**.

The Internal Rate of Return

The internal rate of return is an alternative approach used in making investment decisions, which also takes into account the time value of money. The IRR represents the return (in present terms) earned on an investment over its economic life. It is defined as that interest rate which, when applied to the cash flows generated by an investment, will equate the present value of the cash inflows to the present value of the cash outflows. In other words, it is the discount rate which will cause the NPV of an investment to be zero. The pineapple company investment had a positive NPV of $350 at the discount rate of 10%, implying that the IRR was greater than 10%. Should the NPV amount to a negative figure, then the IRR would be less than 10%. Thus, the zero value of NPV would indicate an IRR at par with the discount rate. The IRR is found by solving the value of **r** in the following formula.

$$C_0 \text{ (investment outlay)} = \frac{CF_1}{(1 + r)} + \frac{CF_2}{(1 + r)^2} + \ldots\ldots + \frac{CF_n}{(1 + r)^n}$$

It is easier, however, to use small hand calculators or computers (for tedious calculations) for a fast and accurate solution. Applying this formula to the pineapple company investment, an IRR of 10.3% is estimated using the hand calculator to solve the following equation:

$$30{,}000 = \frac{5{,}000}{(1 + r)} + \frac{9{,}000}{(1 + r)^2} + \frac{12{,}000}{(1 + r)^3} + \frac{4{,}000}{(1 + r)^4} + \frac{10{,}000}{(1 + r)^5}$$

The mathematics underlying the discount factors are based on the assumption that any cash flows in future years will occur in one lump sum at the year end. However, initial investment outlay is assumed to occur at the beginning of Year 1.

Comparison of NPV and IRR

If taken as the basis for decisions, the IRR method will often result in the same conclusion as the NPV method if taken as the basis for decisions. However, there are also situations where the IRR may lead to different decisions from those indicated by the NPV. Where projects are mutually exclusive (where acceptance of one project excludes the acceptance of another, for example, the choice of one of several possible coffee factory sites), application of the IRR and the NPV rules can lead to conflicting recommendations. This is illustrated by a simple example.

A coffee factory can select one of two mutually exclusive projects. The estimated cash flows are:

	Initial Investment	**Cash Inflow over 3 Years of Project Life**		
	($)	**Year 1**	**Year 2**	**Year 3**
Project X	5,600	2,700	2,700	2,700
Project Y	9,600	4,400	4,400	4,400

The cost of capital is 10%.

The NPV and IRR calculations give the following result:

	NPV ($)	**IRR** (%)
Project X	1,115	21
Project Y	1,342	18

It is seen that the IRR ranks X first, but the NPV ranks Y first. If the projects were independent, this would be immaterial, as both projects would be accepted. But in the context of mutually exclusive projects, the ranking is crucial as only one project can be accepted.

Another problem with the IRR method is that it expresses the result as a percentage rather than in monetary terms. Comparison of percentage returns can be misleading; for example, compare an investment of $50 which produces an IRR of 80% with an investment of $500 producing a return of 20%. Where capital is not a constraint, the investment costing $500 would be chosen as it yields $100 compared to $40 generated by the other investment, although the latter has a superior IRR.

A further problem with the IRR is encountered when unconventional cash flows occur, with negative cash flows coming in later years. If the sign of the net cash flows changes in successive periods, it is possible for the calculations to produce multiple IRRs. While multiple rates of return are theoretically possible, only one rate of return is significant in an 'accept' or 'reject' decision. Analysts should be aware of such situations where the NPV rule gives an unambiguous decision criterion.

The assumption concerning the reinvestment of interim cash flows from the acceptance of projects provides another reason for supporting the superiority of the NPV method. The implicit assumption, when the NPV is applied, is that the interim cash flows will be reinvested at the cost of capital, i.e. the discount rate. However, the IRR rule makes a different implicit assumption about the reinvestment of the interim cash flows. It assumes that all the proceeds from a project can be reinvested to earn a return equal to the IRR of the original project. In the above example, the NPV method assumes that the annual cash inflows of $2,700 for Project X will be reinvested at a cost of capital of 10% whereas the IRR method assumes that they will be reinvested at 21%. In theory, a company will accept all projects which offer a return exceeding the cost of capital, and any other funds which become available can only be reinvested at the cost of capital. This is the assumption implicit in the NPV method.

The conclusion of the above analysis appears to favour the NPV method for investment appraisal because IRR is unreliable in ranking projects where either different outlays are involved or projects are mutually exclusive. This is because the IRR approach assumes that the reinvestment rate is equal to the indicated rate of return over the remaining life of the proposal. The NPV method however implies reinvestment at a rate equivalent to the required rate of return used as the discount rate which approximates to the opportunity rate for investment. The IRR cannot be used if the cost of capital changes during the project. However, firms do use IRR since managers often find it easier to interpret, and the cut-off rate need not be specified beforehand.

IV. BASIC FINANCIAL ACCOUNTS

The discussion so far has shown that financial analysis has various objectives; however, the most fundamental is the provision of guidance to the management of a firm in taking investment decisions. In order to help in making such decisions, discounted cash flow methods of investment appraisal were reviewed, which could assist in deciding whether any particular investment will result in the company maximizing its net cash flows. Two appraisal methods were mentioned - IRR and NPV. Central to both methods is the basic concept of cash flows through time. This is because the increase in the financial resources which the shareholders derive from investments depends upon the amount and timing of the cash flows that can be withdrawn from the business operations. In this connection, simple cash flows have been discussed earlier to introduce some of the measures of return to investment. It is opportune now to introduce the basic accounting principles through examining a set of financial accounts of revenue generating entities which could be used in estimating project cash flows.

This chapter outlines the objectives of financial statements and reviews the information requirements of users of accounts. This is followed by a presentation of a simple set of accounts and explanations as to how these statements relate to each other, and how they are put together in a consistent way. Finally, the broad assumptions and concepts that underlie the construction and presentation of accounts are reviewed.

Objectives of Financial Statements

The fundamental objective of financial statements is to communicate measurements of, and information about, the resources and performance of the reporting entity to those having reasonable rights to such information.

Within this broad framework, an accounting system is intended to: (a) measure the extent to which resources or income have been created by the activities of the enterprise; (b) provide reliable information about changes in the assets and liabilities of an enterprise that result from revenue generating activities; and (c) supply quantitative information that assists in estimating the earning potential of the enterprise.

Financial statements are prepared to present the required information in a summarized form. These statements include balance sheets, profit and loss (or income) statements, source and application of funds statements, and other statements and material which collectively are intended to give a true and fair view of the financial position and results of operations of an enterprise. A true and fair view implies appropriate classification and grouping of the items in the financial statements. It also implies the consistent application of generally accepted accounting principles.

The range of possible information on the activities of companies that could be shown in the reports is so large that for practical reasons priorities have to be selected and accounting practices established which, as far as possible, lead to the disclosure of useful information. Towards this end, the form and content of financial statements are governed by statutes, statements of standard accounting practices, stock exchange regulations and customs and practices within the particular industry. Compliance with these requirements should help achieve the qualitative objectives of financial information, namely **objectivity** (verifiability), **realism** (relevance), **consistency**, **comparability**, **intelligibility**, and **ease and economy of preparation**.

Users' Information Requirements

Company accounts are used for a wide variety of purposes by different groups of users and the accounting profession has a difficult task in trying to meet all such requirements while at the same time keeping accounts as clear and simple as possible. Users of accounts may be classified into certain groups; their requirements for information may not be the same, nor are they likely to regard certain types of information as equally important. Essentially, the task involves achieving an acceptable compromise between different requirements for information.

The following list, while not intended to be exhaustive, contains the main groups of users of company accounts.

(i) **The management.** The depth and detail of accounting information used for management purposes is likely to vary according to the level of management concerned. Senior management and Boards may make more use of the company's published accounts than other levels of management but they also need more frequent and up-to-date information in order to monitor the progress of the company's activities against budgets, and to arrive at decisions which ensure the most efficient use of company's resources. Another major concern of the management is for information on the liquidity position of the company: i.e. forecasts of cash flow during the short and medium term.

(ii) **The investor group.** Existing and potential shareholders will use accounting information to ascertain the security of their investment, and the dividend cover, and to compare these with what can be expected from alternative investments.

(iii) **The loan creditor group.** Creditors and lenders will be primarily concerned with the security of their loan and with the company's ability to meet the agreed repayments as they fall due. They require information on:

(a) company's cash position and its likely requirements for cash in the immediate period ahead to finance the continuation of business; and

(b) the value of the company's net assets.

(iv) **The employee group.** They are primarily interested in: (a) the future of the company, the prospects for expansion and for continuing employment; (b) getting a fair and reasonable share of the company's gains; and (c) the ability of the company in the short-term to meet its requirements for cash, i.e. to remain solvent.

(v) **The analyst-adviser group.** They need particularly detailed information often relating to an extended period to give the best advice to their clients.

(vi) **The business contact group.** Suppliers selling on credit would require information similar to those required by creditors to assess the creditworthiness of the company. Customers use financial information to judge the ability of the company to fulfil its commitments, say a building contract. They may also be able to ascertain fairness of prices charged, reliability and regularity of business conducted.

(vii) **The government.** The government uses accounting information in several ways:

(a) as a customer: to check on excessive profiteering;

(b) as a tax collector: to assess tax liability;

(c) as a regulatory body: to regulate the relationship of businesses with each other and also with the business community, e.g. through price control;

(d) as a central statistical agency: to collect and publish statistical data about different aspects of the business community.

From the above, it can be gleaned that there is a common requirement by most users of accounts for information on a company's net asset values, profitability and liquidity position. However, reporting systems have traditionally placed more emphasis on satisfying the needs of the equity investor and loan creditor group than the needs of the other groups.

Essential Features of the Financial Accounts

In order to illustrate the components of a basic set of financial accounts, a simple example is given below. In essence, the **balance sheet** measures the net assets, i.e. the assets less liabilities at a particular date; and the **profit and loss statement,** the gain or loss over a particular period. Balance sheets show the assets owned by the entity on one side and claims against those assets (by the creditors and owners) on the other side. The equality between the two sides is derived from the basic accounting equation - Assets = Liabilities + Equity. This in turn, is based on the concepts of the business entity (i.e. an entity has an accounting personality of its own independent of its owners) and duality of transactions (i.e. any transaction by the entity has a dual effect).

Recognizing the independent personality of the company, equity is the amount which would eventually be due to shareholders from a company in the case of winding-up. This claim can be satisfied only after all other claims of the company have been liquidated.

In the above context, it must be stressed that a separate personality assumed under the business entity concept has nothing to do with legal personality. The business entity concept is relevant irrespective of the type of ownership - whether sole proprietorship, partnership, limited liability company, corporation or cooperative. The first two lack legal personality in the sense that the owners rather than the businesses run by them are legally liable for all debts incurred, to be recovered out of their personal assets if the business assets are insufficient. In the other cases, legal liability is limited to business assets and to personal assets only to the extent of the amount unpaid on the shares.

An Example

The transactions for the year of 1990 of Y. Worry Ltd., a limited liability company which deals in vegetables, are given below. The effect of each transaction on the balance sheet is to either:

(1) increase one asset and reduce another by an equal amount (e.g. payment of cash for equipment); or

(2) increase one liability and reduce another by an equal amount (e.g. increase bank overdraft to pay suppliers); or

(3) increase an asset and increase a liability by an equal amount (e.g. buying vegetables on credit); or

(4) reduce an asset and reduce a liability by an equal amount (e.g. payment of cash to creditors).

The content of the balance sheet changes after each transaction and a balance sheet can be prepared after every transaction.

Transactions	**... $ ...**
1. Share capital received	10,000
2. Purchased equipment for cash	3,000
3. Cash purchase of vegetables	4,000
4. Credit purchase of vegetables	2,000
5. Credit sale of 75% of vegetables	10,000
6. Paid wages and other expenses	2,000
7. Received payment from debtors	5,000
8. Paid creditors in full	2,000
9. Wages incurred but not paid	200

Note: (a) Stock remaining unsold is valued at 1,500

(b) The life of the equipment is estimated at 10 years.

The dual aspect of these transactions is demonstrated below:

Transaction	Assets		Liabilities	
	Increase	Decrease	Increase	Decrease
1. Share capital received	Cash		Equity 1/ (share capital)	
2. Equipment purchase for cash	Equipment	Cash		
3. Cash purchase of vegetables	Stock (vegetables)	Cash		
4. Credit purchase	Stock (vegetables)		Creditors	
5. Sale of vegetables	Debtors	Vegetables		
6. Paid wages and expenses		Cash		Equity 1/ (P & L a/c)
7. Receipts from debtors	Cash	Debtors		
8. Payment to creditors		Cash		Creditors
9. Accrued wages (Expense creditors)			Accrued wages	Equity 1/ (P & L a/c)

1/ Equity is ordinary share capital plus reserves including profit and loss balance.

The effect of each transaction may also be shown in a series of balance sheets as follows.

Transaction	**1**	**2**	**3**	**4**	**5**	**6**	**7**	**8**	**9**
Number (Date)	**(1/1/90)**								**31/12/90**
	($)	**($)**	**($)**	**($)**	**($)**	**($)**	**($)**	**($)**	**($)**
Liabilities									
Share capital	10,000	10,000	10,000	10,000	10,000	10,000	10,000	10,000	10,000
P & L account					5,500	3,500	3,500	3,500	3,300
Equity	**10,000**	**10,000**	**10,000**	**10,000**	**15,500**	**13,500**	**13,500**	**13,500**	**13,300**
Creditors				2,000	2,000	2,000	2,000		
Expense creditors									200
Total	**10,000**	**10,000**	**10,000**	**12,000**	**17,500**	**15,500**	**15,500**	**13,500**	**13,500**
Assets									
Equipment		3,000	3,000	3,000	3,000	3,000	3,000	3,000	3,000
Stock (veg.)			4,000	6,000	1,500	1,500	1,500	1,500	1,500
Debtors					10,000	10,000	5,000	5,000	5,000
Cash	10,000	7,000	3,000	3,000	3,000	1,000	6,000	4,000	4,000
Total	**10,000**	**10,000**	**10,000**	**12,000**	**17,500**	**15,500**	**15,500**	**13,500**	**13,500**

Y. Worry Ltd.

Balance Sheet as at 1.1.90

Equity & Liabilities	**($)**	**Assets**	**($)**
Share capital	10,000	Cash	10,000

Y. Worry Ltd.

Balance Sheet as at 31.12.90

Equity & Liabilities	**($)**	**Assets**	**($)**
Share capital	10,000	Equipment	3,000
P & L balance	3,300	Stock	1,500
Equity	**13,300**	Debtors	5,000
Trade creditors	-	Cash	4,000
Expense creditors	200		
Total Equity & Liabilities	**13,500**	**Total Assets**	**13,500**

The Measurement of Profit

The effect of transaction 5, where there is a sale at a profit of $5,500, is to increase total assets by the amount of this profit. This amount belongs to the shareholders and is shown as part of their equity, thereby increasing their claim on the assets of the business. Similarly, the effect of transaction 6, where wages and expenses of $2,000 are paid, is to reduce the total assets by this amount. This expense reduces the owners' claim on the assets of the business. In addition, transaction 9 results in an expense of $200 for services rendered by workers, although not paid yet. This has a similar effect as transaction 6. In short, profits or gains which increase the total assets equally increase equity. Expenses or losses reduce the total assets and are matched by a reduction in equity.

The profit or loss for any period may be determined by deducting opening from closing equity, subject to the necessary adjustments for any capital redeemed or any additional capital paid in by owners during the same period. In the above example, profit for the period is $13,300 - $10,000 = $3,300. This method of arriving at profit, however, is not sufficiently informative to allow management to make meaningful decisions.

A profit and loss statement comparing revenue and related expenses is more useful, as can be seen from the presentation below:

Y. Worry Ltd.
Profit and Loss Statement
for the Year ended 31.12.90

	($)	($)
Sales		10,000
Less cost of sales		
Opening stock	-	
Purchases	6,000	
Less closing stock	1,500	4,500
Gross Profit		**5,500**
Less wages and expenses		2,200
Net Profit		**3,300**

Observations

Several things can be noticed from the above statements.

(1) First thing to note is the date. A balance sheet reports assets and liabilities of the enterprise at a moment in time. A profit and loss statement does not refer to a moment in time, but to a period such as a year from 1.1.90 to 31.12.90, and reports in summary fashion the flow of resources through the enterprise in the course of its operations. The balance sheet thus measures a stock; while the profit and loss statement measures a flow.

(2) The presentation of the statement can be in conventional form like the balance sheet listing assets on one side and liabilities on the other, or in vertical form like the profit and loss statement. However, for ease of understanding by the layman, vertical presentation is increasingly preferred in published accounts.

(3) Every change in the balance sheet between two dates can be accounted for by events that occurred during the year.

(4) Some transactions affect the balance sheet but have no effect on the profit and loss statement. For example, the receipt of share capital contributions increased the share capital and cash figures in the balance sheet. Similarly settlement of creditors eliminated the creditors balance and reduced the cash balance, both in the balance sheet.

(5) Net profit from operations increased the owners' equity since it was retained and not distributed during the year 1990.

(6) Although the company made a net profit of $3,300, the cash balance decreased by $6,000, falling from $10,000 at the beginning of the year to $4,000 at the end of the year. Profitability must be distinguished from liquidity. The difference can be explained by drawing up a statement of source and application of funds. There are several ways of presenting this statement. As presented here, it seeks to reconcile the net profit to cash changes, comparing the opening and closing balance sheet.

Y. Worry Ltd.
Statement of Source and Application of Funds for the Year ended 31.12.90

	..... $	
Sources of Funds		
- Net profit		3,300
Add **other sources of funds**		
- Increase in creditors		200
Total source		**3,500**
Less **application of funds**		
- Purchase of equipment	3,000	
- Increase in stock	1,500	
- Increase in debtors	5,000	9,500
Decrease in cash balance		**6,000**

(7) The profit and loss statement covering a year's operation provides a link between the opening balance sheet and closing balance sheet, i.e. opening net assets (opening assets - opening liabilities) amount to $10,000, which, added to profit for the year of $3,300, gives the closing net assets at $13,300 (opening assets - closing liabilities). A more general relationship between these two statements is as follows.

Relationship Between Balance Sheet and Profit and Loss Account

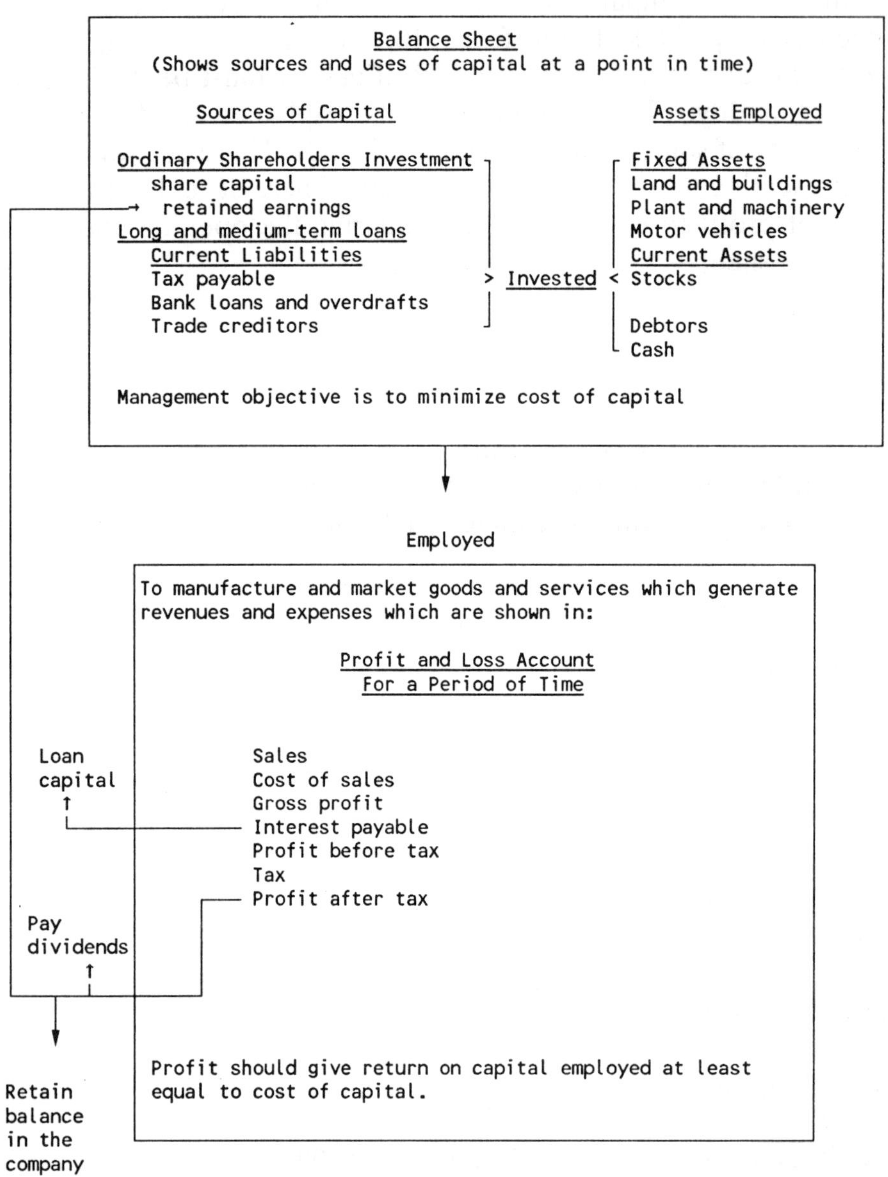

(8) The statement of source and application of funds is prepared from financial data generally identifiable in the profit and loss statement, the balance sheet and related notes. But it presents information which is not readily available in a usable form in the other statements.

Although the profit and loss statement is the starting point for the funds statement, the two statements differ in their treatment of several important accounting entries. The funds statement gives a more complete accounting of debt transactions by showing principal payments and proceeds of new loans whereas the profit and loss statement shows only interest charges. The funds statement reflects the cash transactions that occur with the purchase and/or sale of capital assets. On the profit and loss statement, expenses associated with capital assets are determined by the process of depreciation which represents a portion of the value of the asset consumed in generating the sale.

The relationship among these three statements is presented below.

Coordination among Financial Statements of Y. Worry Ltd.

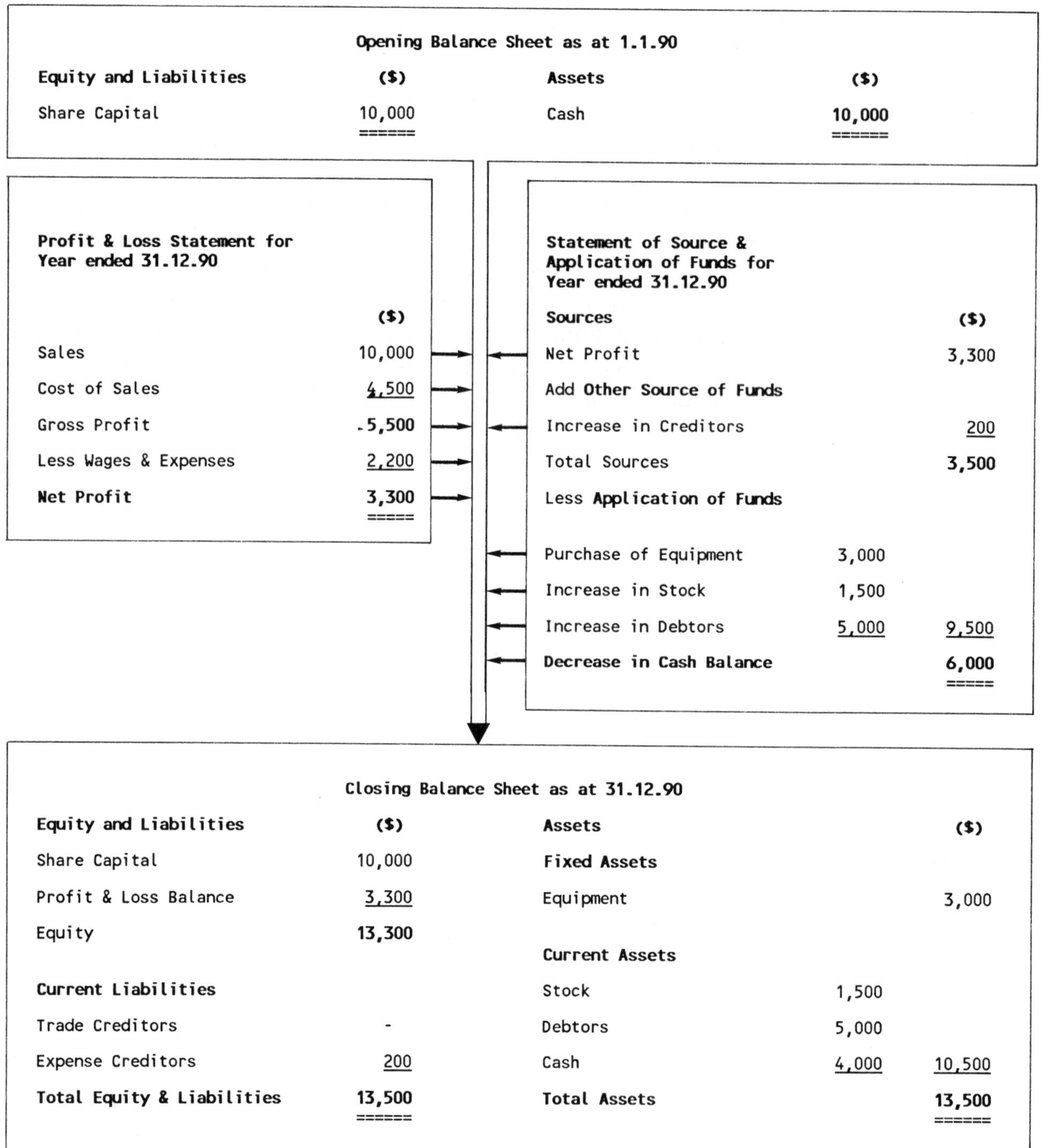

Opening Balance Sheet as at 1.1.90

Equity and Liabilities	($)	Assets	($)
Share Capital	10,000	Cash	10,000

Profit & Loss Statement for Year ended 31.12.90

	($)
Sales	10,000
Cost of Sales	4,500
Gross Profit	5,500
Less Wages & Expenses	2,200
Net Profit	**3,300**

Statement of Source & Application of Funds for Year ended 31.12.90

Sources		($)
Net Profit		3,300
Add **Other Source of Funds**		
Increase in Creditors		200
Total Sources		3,500
Less **Application of Funds**		
Purchase of Equipment	3,000	
Increase in Stock	1,500	
Increase in Debtors	5,000	9,500
Decrease in Cash Balance		**6,000**

Closing Balance Sheet as at 31.12.90

Equity and Liabilities	($)	Assets		($)
Share Capital	10,000	**Fixed Assets**		
Profit & Loss Balance	3,300	Equipment		3,000
Equity	13,300			
		Current Assets		
Current Liabilities		Stock	1,500	
Trade Creditors	-	Debtors	5,000	
Expense Creditors	200	Cash	4,000	10,500
Total Equity & Liabilities	**13,500**	**Total Assets**		**13,500**

(9) Finally, as shown in the recording of transactions, every item in the balance sheet has come about as a result of cash flows. Also the size of every item has been determined by these cash flows. The equity and the long-term liabilities represent a flow of cash which has come into the business over the years while the current liabilities comprise claims, the discharge of which will cause movements of cash out of

the business in the near future. Similarly fixed assets have been acquired as a result of cash outlays. The success of a business depends on the balance between these cash inflows and outflows. If the latter are consistently less than the former, the business will survive and possibly flourish. If the reverse is true, it will shrink and eventually fold up. As with the business as a whole, so with the individual assets, which make up the enterprise; the contribution of each asset to the health of the entity depends on the balance of the cash flows associated with it. Hence in considering the attractiveness or otherwise of an investment project either by itself, or as part of a project or enterprise, attention must be focused on cash flows.

Accounting Principles

The broad assumptions and concepts that underlie the construction of accounts and their presentation are reviewed in this chapter. These are fundamental to the understanding and interpretation of the financial statements.

Accounting Concepts

The money measurement concept. The accounting system only records those facts and events that can be expressed in monetary terms. This concept imposes severe limitations on the scope of accounting statements. The accounts of a company do not reveal, for example, that a competitor has introduced a new product which is technologically superior to the company's product. Moreover, while money is probably the only practical denominator, the use of money implies homogeneity, a basic similarity between $1 and another. In periods of inflation, this homogeneity does not in fact exist. Thus, transactions recorded in the books of Y. Worry Ltd. carry monetary values. Where values cannot be applied such as for the skill of sales staff, they are not reflected.

The business entity concept. Accounts are maintained for business entities as distinct from the persons who own them, operate them or are otherwise associated with the business. The balance sheet and profit and loss statement shown earlier are concerned with Y. Worry Ltd., a company which is distinct from Mr. Y. Worry, who may be one of the shareholders, albeit, the largest shareholder.

The going concern concept. While preparing financial statements, it is assumed that the business will continue to operate for the foreseeable future. Accounts do not attempt to measure at all times what a business is currently worth to a potential buyer. Nor do they show the value of the assets of the company if it went into voluntary liquidation. Unless the company has recently revalued its assets, they are shown in a historical cost accounting system not at their current value but rather at their original cost. The current value may be possibly above or below the actual cost shown in the accounting records. The value of the assets of Y. Worry Ltd. were recorded at their original cost (cost at which they were purchased) which may realize less if some of the assets - debtors or stock contain irrecoverable items, (for example, bad debts or spoilt vegetables).

The accruals concept. In accounting, revenues and costs are recognized (or accrue to the business) as they are earned or incurred, not as money is received or paid. Revenue and costs are matched with one another so far as their relationship can be established or justifiably assumed, and dealt with in the income statement of the period to which they relate. For example, the wages and expenses incurred to generate sales were $2,200, properly chargeable to the profit and loss account, not the cash payment of $2,000.

The consistency concept. There should be consistency in the accounting treatment of like items within each accounting period, and in the accruals of successive periods. If there is any departure from consistent treatment of like items, this must be justified and noted. For example, stock is valued at cost which is $1,500 (25% of $6,000) in 1990. If closing stock is valued at cost plus some overhead in 1991, then profit figures of 1990 and 1991 cannot be compared. It does not mean that the valuation should be the same as before but reasons for deviation must be noted.

The concept of prudence. Revenues and profits are never anticipated and only included in financial statements when they are realized; however, provision is made for all known liabilities (expenses or losses). Depreciation on equipment (a known expense of $300 for Y. Worry Ltd.) has not been charged although the equipment is stated to have a life of only 10 years (assuming straight line method and 'nil' residual value). This omission overstates the profit by a similar amount and contravenes the concept of prudence.

The cost concept. A fundamental concept of accounting closely related to the money measurement and going concern concepts is that an asset is ordinarily entered in the accounting records at the price paid to purchase the asset, and that cost is the basis of all subsequent accounting for the asset. The accounting measurement does not normally reflect the worth of assets except at the moment they are acquired. This situation may result, particularly in a period of rapid inflation, in the production of accounting statements which make it difficult to measure how efficiently the company's management have carried out their stewardship function.

The cost concept does not imply that all assets remain in the accounting records at their original cost for as long as the company owns them. The cost of fixed assets is systematically reduced over the life of the asset by the process of depreciation. The depreciation is charged to the profit and loss account representing the portion of the asset's cost utilized during the accounting period. It is important to appreciate that the depreciation process does not provide a fund to replace the asset at the end of its useful life. It does reduce the profit available for distribution and, therefore, the potential outflow of cash from the business in the form of dividends to shareholders. The omission of depreciation on equipment by Y. Worry Ltd. has been noted under concept of prudence. This omission contravenes the cost concept as well.

Another important consequence of the cost concept is that if the company pays nothing for an asset it acquires, this asset will normally not appear in the accounting records as an asset. The knowledge, skill and expertise of a company's staff do not appear in the balance sheet as an asset. The failure of accounting statements to reflect the value of human assets can give rise to significant differences between the book value, the break-up value, the value of the company if sold as a going concern, and the market value of a company's shares.

The concept of realization. An example helps in the exposition of the realization principle. Suppose an enterprise produces a power-tiller in one accounting period, stores it through a second, sells it in the third and collects the cash in the fourth. If the cost is $4,000 and selling price is $5,000, the profit is obviously $1,000. The problem arises because of the accounting period assumption. Accounts need to be produced for each of the four periods but how is the profit divided up between the four periods? Obviously a number of possibilities exist, but by convention the whole of the profit is recognized in the period in which the sale is made, i.e. the period in which the revenue (but not necessarily in the form of cash) is realized. In this example, it is the third period.

Accounting Bases and Policies

Difficulties can and do arise in applying the fundamental accounting concepts. The main difficulty stems from the fact that many business transactions have financial effects spreading over a number of years. Accountants have to make decisions on the extent to which expenditure incurred in one year can reasonably be expected to produce benefits in the form of revenue in other years and whether these should be carried forward in whole or in part; in other words, to what extent expenditure should appear as a charge in the income statement of the current year as distinct from appearing in the balance sheet as a resource at the end of the year. All decisions require consideration of future events and an element of commercial judgement is unavoidable in the assessment. Significant matters which require judgement include depreciation of assets, valuation of stocks and work in progress, repairs and renewals, research and development expenditure and long-term contracts.

Accounting bases are the methods which have been developed for expressing or applying fundamental accounting concepts to financial transactions and items. **Accounting policies** are the specific accounting bases judged by the business enterprises to be most appropriate to their circumstances and adopted by them for the purpose of preparing financial accounts. Thus depreciation is charged in accordance with the accrual concept. But there can be several bases for making the charge - straight line, diminishing balance, production unit, production hours, or sum of the years' digit methods. The particular method selected (i.e. straight line method) becomes part of the accounting policy for the enterprise and must be applied consistently. Two considerations should govern the selection of such policies:

(a) **Substance over form.** Transactions and other events should be accounted for and presented in accordance with financial reality and not merely with their legal form. For example, an item of equipment under hire-purchase will be recorded as an asset in the balance sheet although legally the enterprise would not have become the owner until a certain number of instalments have been paid.

(b) **Materiality.** Financial statements should disclose all items which are material enough to affect evaluations or decisions. For example, an enterprise may decide to charge to the profit and loss account capital expenditure on, say, staplers of $24 (with life of 3 years) in the year of purchase, although, in principle, only annual depreciation of $8 should be charged over 3 years. This is because the expenditure is very small compared to other capital expenditures and the cost of maintaining a depreciation schedule and records will be excessive in relation to the expected benefit from maintaining such records.

The users cannot make reliable judgements unless the financial statements clearly disclose the significant accounting policies which have been adopted in preparing them. The following are examples of areas in which differing accounting policies exist and which therefore require disclosure of the treatment selected.

General. Consolidation policy (i.e. for a company which has subsidiaries), conversion or translation of foreign currencies including the treatment of exchange gains and losses, overall valuation policies (e.g. historical cost, general or current purchasing power, replacement cost and current cost), leases, hire purchase or instalment transactions, taxation and long-term contracts.

Assets. Accounts receivables, inventories (raw materials, finished stock and work in progress), depreciable assets, depreciation, growing crops, investments, subsidiary companies, associated companies and other investments, research and development, patents, trademarks and goodwill.

Liabilities and provision. Commitments, contingencies, warranties, pension costs, severance and redundancy payments.

Profits and losses. Methods of revenue recognition, maintenance, repairs and improvements, gains and losses on disposal of property and reserve accounting.

An Illustration

An illustration is provided to show the effect of differing accounting policies. Consider a simplified example of two sugar factories which are identical in every respect other than the accounting policies they adopt for stock valuation and depreciation.

	Company X		Company Y	
Year 1		.... ($)		
Sales		3,750		3,750
Less cost of sales		2,450		2,900
Gross Profit		**1,300**		**850**
Depreciation	150		225	
Other expenses	400	550	400	625
Net Profit		**750**		**225**
Year 2				
Sales		4,500		4,500
Less cost of sales		3,050		3,600
Gross Profit		**1,450**		**900**
Depreciation	150		190	
Other expenses	450	600	450	640
Net Profit		**850**		**260**

Company X values its stock inclusive of both direct and indirect costs of production and depreciates its fixed assets at 10% per annum on a straight line basis; whereas Company Y values its stock at direct production cost only and depreciates its fixed assets at 15% per annum on a reducing balance basis. On the assumption that each company has identical physical asset structures and accountable transactions, the profit statements could be as shown above.

From the same basic data, therefore, these two companies, for no other reason than the different accounting policies adopted, show totally different profit figures for the two years. Fixed asset and stock valuation in the balance sheets would also be different. Despite the adoption of different accounting policies, the total reported profits of each of these companies would be the same over their whole lives from incorporation to liquidation. It is therefore the different accounting policies used to apportion these total profits to designated accounting periods which cause problems. The danger is that the differences in reported profits could result in significantly different decisions by users of accounting information. The danger can be minimized by disclosing the accounting policies adopted in drawing up the accounts.

V. FINANCIAL COST AND BENEFIT CASH FLOWS

In Chapter III, it is argued that financial cash flows are the heart of DCF calculations and that the IRR (and NPV) calculation calls for the estimation of costs and revenues attributable to the project. The present chapter develops a methodology to estimate, on an annual basis, the **relevant cash outflows and cash inflows.** Annual balance sheet, profit and loss and sources and application of funds statements are projected and, based on these projections, incremental cash flow streams before and after financing are derived and the relevant IRRs calculated. Projections are made in constant terms (as of a particular date). The effects of price changes on debt service are also noted.

A Comprehensive Illustration

New Project

A simplified example is given below. It refers to a new company - Beta Ltd. - incorporated in 1990 to undertake a vegetable oil milling business. The bases for estimates of costs and revenues are:

(1) Investment costs in the first year (1990) are:

	. . $ . .
- Land	100,000
- Buildings	200,000
- Plant, machinery and equipment	500,000
Total	**800,000**

(2) **Plant capacity:** processing of 5,000 tons of oilseed/year to produce, from Year 3:

Oil 15%	=	750 ton/year
Meal 80%	-	4,000 ton/year
Processing losses	:	5%

In Year 2, about 90% of capacity is expected to be used. In Year 1 the plant would be installed and no use would be made.

(3) **Sales prices:** US$500/ton oil; US$300/ton meal.

(4) **Raw material prices:** US$200/ton oilseed.

(5) **Packing material:** US$50/ton of oil processed.

(6) **Labour:** total wages per year are estimated at US$100,000. In Year 2, full wages would be paid.

(7) **Fuel use:** 0.1 ton fuel per 1 ton oilseed processed. Fuel price US$300/ton.

(8) **Overheads:** administrative and marketing costs are estimated at US$20,000/year.

(9) **Depreciation rates:** buildings 3% p.a.
plant machinery and equipment: 10% p.a.

(10) **Project life:** assumed at 6 years.

(11) **Working capital requirements:**

Cash & bank: US$50,000
Receivables 30 days' sales (oil and meal)
Inventories 30 days' requirements (oilseeds & packing material).

(12) **Interest:** on long-term loans 10% p.a
on short-term loans 14% p.a.

The following schedules and analyses can be prepared:

(1) Capital investment and depreciation schedule (Schedule 1).

(2) Estimated production and revenues (Schedule 2).

(3) Estimated inputs and costs (Schedule 3).

(4) Estimated working capital requirements (Schedule 4).

(5) Projected profit and loss account (Schedule 5).

(6) Projected source and application of funds statement (Schedule 6).

(7) Projected balance sheet (Schedule 7).

(8) Financial rate of return analysis (Schedule 8).

SCHEDULE 1

Capital Investment and Depreciation

Item	Investment	Rate	... Depreciation ... Annual Depreciation
	($'000)	(%)	($'000)
Land	100	-	-
Buildings	200	3	6
Plant, machinery and equipment	500	10	50
Total	**800**	-	**56**

SCHEDULE 2

Estimated Production and Revenues ($'000)

	Year 1	Year 2	Year 3	Year 4	Year 5
Production					
- Oil (tons)	-	675	750	750	750
- Meal (tons)	-	3,600	4,000	4,000	4,000
Sale Prices					
- Oil ($/ton)	500	500	500	500	500
- Meal ($/ton)	300	300	300	300	300
Revenue ($'000)					
- Oil sales	-	338	375	375	375
- Meal sales	-	1,080	1,200	1,200	1,200
Total	-	**1,418**	**1,575**	**1,575**	**1,575**

This schedule matches sales with production; there are no closing stocks of either work in progress or finished goods.

SCHEDULE 3

Estimated Inputs and Costs ($'000)

	Year 1	Year 2	Year 3	Year 4	Year 5
Inputs Used					
- Raw material - oilseeds (tons used)	-	4,500	5,000	5,000	5,000
- Fuel oil (tons used)	-	450	500	500	500
Cost of Sales ($'000)					
- Raw material	-	900	1,000	1,000	1,000
- Packing material	-	34	38	38	38
- Labour		100	100	100	100
- Fuel oil	-	135	150	150	150
Total	-	**1,169**	**1,288**	**1,288**	**1,288**

The above schedule presents cost estimates for each period. The costs are based on the volume of production expected during that period and not on the volume of sales projected for that period. Thus cost of raw material, packing material, labour and fuel oil refer to volumes produced. In this illustration, since both sales and production volumes are the same, costs on both bases would be the same. It is worth noting that all raw materials (and packing materials) purchased were not used and hence there remain closing balances which are adjusted through the working capital schedule (see below).

SCHEDULE 4

Working Capital Requirements ($'000)

	Year 1	Year 2	Year 3	Year 4	Year 5
Current Assets					
- Cash and bank (working balances)	-	50	50	50	50
- Receivables (30 days)	-	118	131	131	131
- Inventories (30 days)	-	78	87	87	87
Total Current Assets		**246**	**268**	**268**	**268**
Current Liabilities					
- Payables (15 days)	-	39	43	43	43
Working Capital	-	**207**	**225**	**225**	**225**
Incremental Working Capital	-	**207**	**18**	-	-

In the above schedule, adjustments are made for changes in cash tied up in working capital. The increase in working capital balances is the amount estimated as necessary to support the operations resulting from the project investment. Excess cash has not been included. Instead this is invested in short-term securities (see Schedule 6).

SCHEDULE 5

Profit and Loss Account Projection ($'000)

	Year 1	Year 2	Year 3	Year 4	Year 5	Year 6
Sales revenue	-	1,418	1,575	1,575	1,575	1,575
Cost of sales 1/	-	1,225	1,344	1,344	1,344	1,344
Gross profit	-	193	231	231	231	231
Administrative & marketing costs	-	20	20	20	20	20
Profit before interest	-	173	211	211	211	211
Interest - short-term	-	4	4	-	-	-
Interest - long-term	-	50	50	33	16	-
Net profit	-	119	157	178	195	211
Dividends distribution	-	-	-	66	82	211
Retained earnings	-	119	157	112	113	-
Accumulated retained earnings	-	119	276	388	501	501

1/ Includes depreciation of $56,000 p.a.

Based on the first three schedules, the profit and loss account has been projected. Combined with this and Schedule 4, the balance sheet has been prepared (Schedule 7). These two statements, together with the relevant schedules, provide the base for the projection of funds statement shown in Schedule 6.

SCHEDULE 6

Sources and Application of Funds Projection ($'000)

	Year 1	Year 2	Year 3	Year 4	Year 5	Year 6
Sources of Funds						
Net profit	-	119	157	178	195	211
Depreciation	-	56	56	56	56	56
Funds from operation	-	175	213	234	251	267
Equity investment	300	-	-	-	-	-
L.T. loans disbursement	500	-	-	-	-	-
Total Sources	**800**	**175**	**213**	**234**	**251**	**267**
Application of Funds						
Purchase of land, building & plant	800	-	-	-	-	-
Repayment L.T. debt	-	-	163	168	169	-
Dividends distribution	-	-	-	66	82	211
Incr. (dec) working capital (see below)	-	175	50	-	-	56
Total Application	**800**	**175**	**213**	**234**	**251**	**267**
Changes in Working Capital						
Current Assets						
Increase in cash & banks	-	50	-	-	-	-
Increase in securities	-	-	-	-	-	56
Increase in receivables	-	118	13	-	-	-
Increase in inventories	-	78	9	-	-	-
Total Increase in Current Assets	**-**	**246**	**22**	**-**	**-**	**56**
Current Liabilities						
Increase in short-term debt	-	32	(32)	-	-	-
Increase in payables	-	39	4	-	-	-
Total Increase in Current Liabilities	**-**	**71**	**(28)**	**-**	**-**	**-**
Net changes in Working Capital	**-**	**175**	**50**	**-**	**-**	**56**

A significant point to note is that **funds from operation** (net profit plus depreciation) in Year 2 ($175,000) are able to meet the total working capital ($175,000) to be financed. The company can also meet its debt service in every year, with increasing ease as it accumulates a growing positive cash position.

SCHEDULE 7

Balance Sheet Projections ($'000)

Assets	**Year 1**	**Year 2**	**Year 3**	**Year 4**	**Year 5**	**Year 6**
Fixed Assets						
Land	100	100	100	100	100	100
Buildings	200	200	200	200	200	200
Plant	500	500	500	500	500	500
Less accumulated depreciation	-	56	112	168	224	280
Net fixed assets	**800**	**744**	**688**	**632**	**576**	**520**
Current Assets						
Inventories	-	78	87	87	87	87
Receivables	-	118	131	131	131	131
Securities	-	-	-	-	-	56
Cash and bank	-	50	50	50	50	50
Total Assets	**800**	**990**	**956**	**900**	**844**	**844**
Equity and Liabilities						
Share capital	300	300	300	300	300	300
Retained earnings	-	119	276	388	501	501
Total Equity	**300**	**419**	**576**	**688**	**801**	**801**
Long-term loans	500	500	337	169	-	-
Current Liabilities						
Short-term loans	-	32	-	-	-	-
Payables	-	39	43	43	43	43
Total Current Liabilities	-	**71**	**43**	**43**	**43**	**43**
Total Liabilities and Equity	**800**	**990**	**956**	**900**	**844**	**844**

SCHEDULE 8

Financial Rate of Return Analysis (Pre-Tax)
(in Constant $'000)

Year	**Investment Cost**	**Cash Flows from Operations**	**Net Cash Flow**
1990	800	-	(800)
1991	207 1/	225	18
1992	18 1/	263	245
1993	-	267	267
1994	-	267	267
1995	(225) (150) 2/	267	642

Financial Rate of Return - 17%

1/ Incremental working capital from Schedule 4, recovered in 1995.
2/ Residual value of fixed assets.

Financial rate of return before financing. The above schedule summarizes all the information on cash flows. Investment costs have been taken from Schedule 1. Since it is assumed that the fixed assets will have a residual value of $150,000, this has been included in the last year of the project life along with incremental working capital which is also recovered in full in that year. Investment in incremental working capital is taken from Schedule 4.

It is worth noting that cash flow from operations extracted from Schedule 6 does not include long-term interest or dividends. Long-term interest is excluded to **avoid double counting** while dividends are ignored to enable the IRR to reflect **returns to all resources** rather than only return to equity. Based on the resulting stream of net cash flows, an IRR of 17% is calculated. This represents the financial rate of return to all resources employed.

Financial rate of return after financing. However, investors are concerned about the return on capital after payment of taxes and debt service. These are therefore deducted from the net cash flows before financing to arrive at net cash flows after taxes and financing, which are then discounted. The resulting IRR is not only influenced by the IRR to all resources employed, but also by the terms on which loans are obtained.

For Beta Ltd., the return to funds put in by the shareholders would be 8% computed as follows.

Financial Rate of Return after Tax and Financing

	1989	**1990**	**1991**	**1992**	**1993**	**1994**
Investment cost	800	-	-	-	-	(150)
Investment working capital	-	207	18	-	-	(225)
Loan receipts	500	-	-	-	-	-
Funds from operations after tax 1/ 2/	-	225	203	188	178	169
Debt service:						
- Principal	-	-	163	168	169	-
- Interest	-	50	50	33	16	-
Net Cash Flow	**(300)**	**(32)**	**(28)**	**(13)**	**(7)**	**544**

Return to Equity = 8%

1/ Taken from Schedule 6 and adjusted for long-term interest as the latter is shown separately.
2/ See Chapter VII for tax computation: tax payment 1990 = $60,000, 1991 = $79,000, 1992 = $89,000 & 1993 = $105,000.

Additions to Existing Operations

When the project is merely an addition to the ongoing operations of an enterprise, identification of cash flows relevant to the new project becomes difficult. It becomes necessary to measure the difference between what would occur if the investment under question were undertaken and if it were not undertaken; in effect, the objective is to measure the incremental effect on the firm's overall operations from undertaking the proposed investment. Because every estimate of cash flows involves a comparison (between the projected "with" and "without" scenarios), it is important that the most plausible alternative be projected as the one likely to occur if the investment is not undertaken. A project may appear attractive if the relevant cash flow analysis is made by comparing its performance against an unrealistic and unprofitable alternative. Let us illustrate through a simple example how incremental cash flows are estimated adopting a **with and without** project approach.

Alpha Ltd., a feedmill, currently operates at 60% capacity due to machinery bottlenecks. The company plans to import the required machinery at a cost of $250,000, expects to raise capacity to 80% and reduce variable costs by 10%. An alternative to imported machinery is a locally manufactured substitute costing $80,000. It would raise capacity to 70%, but increase variable operating costs by 4% due to higher consumption of fuel oil. Current sales revenues are $300,000 per annum and it is expected that the selling price will not change from the current level. Variable and fixed costs at the current (60%) capacity are respectively 75% of sales revenue and $60,000 per annum. It is assumed that both the local and imported machinery will have a life of 5 years with 'nil' residual value.

Based on the above information, three schedules are prepared: Schedule 1 shows cash flows with the project (involving imported machinery); Schedule 2 shows cash flows without project (i.e with local machinery); and Schedule 3 shows the incremental cash flows. These schedules are derived from projected profit and loss statements with project and without project. These are shown in Figure 1 and Figure 2. The results indicate that the new investment is very attractive, yielding an internal rate of return of 35%.

SCHEDULE 1

Cash Flows with the Project (in Constant $'000)

	Year 1	Year 2	Year 3	Year 4	Year 5
Capital costs	250	-	-	-	-
Operating costs	330	330	330	330	330
Total Costs	**580**	**330**	**330**	**330**	**330**
Sales Revenue	400	400	400	400	400
Net Cash Flow	**(180)**	**70**	**70**	**70**	**70**

SCHEDULE 2

Cash Flows without Project (in Constant $'000)

	Year 1	Year 2	Year 3	Year 4	Year 5
Capital costs	80	-	-	-	-
Operating costs	333	333	333	333	333
Total Costs	**413**	**333**	**333**	**333**	**333**
Sales Revenue	350	350	350	350	350
Net Cash Flow	**(63)**	**17**	**17**	**17**	**17**

SCHEDULE 3

Incremental Cash Flows

	Year 1	Year 2	Year 3	Year 4	Year 5
Incremental Cash Flows	(117)	53	53	53	53

Financial Rate of Return = 35%

FIGURE 1

Projected Profit and Loss Statement with Project (in Constant $'000)

	Year 1	Year 2	Year 3	Year 4	Year 5
		.... (%)			
Capacity utilization	80	80	80	80	80
		.... ($'000)			
Sales revenue	400	400	400	400	400
Less operating costs	330	330	330	330	330
- Variable 1/	270	270	270	270	270
- Fixed	60	60	60	60	60
Cash Operating Profit	**70**	**70**	**70**	**70**	**70**

1/ $75\% \times 400{,}000 \times \frac{90}{100} = \$270{,}000$

FIGURE 2

Projected Profit and Loss Statement without Project (in Constant $'000)

	Year 1	Year 2	Year 3	Year 4	Year 5
			.... (%)		
Capacity utilization	70	70	70	70	70
			.... ($'000)		
Sales revenue	350	350	350	350	350
Less operating costs	333	333	333	333	333
- Variable 1/	273	273	273	273	273
- Fixed	60	60	60	60	60
Cash Operating Profit	**17**	**17**	**17**	**17**	**17**

1/ $75\% \times 350{,}000 \times \frac{104}{100} = \$273{,}000$

Inflation and Debt Service

The effect of **inflation on debt service** in the context of project appraisal is examined in the following example.

A rice dealer plans to construct storage capacity costing $50,000. This is expected to enable him to earn additional revenue of $8,000 a year, with extra costs of $2,000 for 20 years (the life of the store). A development bank will advance him a loan of $50,000 at 10% for 20 years. Construction will be completed by the end of Year 1.

FIGURE 1

Rice Storage Project: IRR

End of Year	Investment Cost	Operating Cost	Revenue	Net Cash Flow
		.... ($'000)		
0	50	-	-	(50)
1-20	-	2	8	6

IRR before Debt Service = 10.5%

The annual cash flow of the project at $6,000 is just sufficient to service the bank loan. However, the project is sensitive to quite small variations in revenue which could occur from **smaller paddy harvests, lower prices** or **simply forecasting errors.** For example, if revenue is 40% lower than the estimate, the surplus drops to $2,800 (60% of $8,000 - $2,000) and the project will be unable to service its debt out of current revenue.

Now let us explore the effect of inflation of 15% a year, affecting costs and revenues (but not debt service) equally. Figure 3 demonstrates how the project becomes more profitable in real as well as in money terms.

FIGURE 2

Rice Storage Project: Cash Flow without Inflation

End of Year	Investment Cost	Operating Cost	Revenue	Surplus	Loan	Net Cash Flow
		($'000)				
0	(50)	-	-	-	50	-
1-20	-	2	8	6	(5.97)	0.03

FIGURE 3

Cash Flow of Rice Storage Project with 15% Inflation

Price Index	End of Year	Cash Flow before Debt Service	Debt Service	Net Cash Flow in Money Terms	Net Cash Flow in Real Terms
		($'000)			
100	0	-	-	-	-
115	1	6.90	5.97	0.93	0.81
132	2	7.92	5.97	1.95	1.48
152	3	9.12	5.97	3.15	2.07
175	4	10.50	5.97	4.53	2.59
201	5	12.06	5.97	6.09	3.03
936	15	56.16	5.97	50.19	5.36
1,883	20	112.98	5.97	107.01	5.68

Figure 4 shows in addition that inflation reduces the project's vulnerability to a drop in revenue - so much so that the project would have a positive cash flow even with 40% lower than expected revenue except in Years 1, 2 and 3. However, even with 15% annual inflation, the project remains vulnerable in the early years to a drop in revenue, particularly as in the short run there is much less certainty that rice prices will rise in line with inflation than in the long run.

FIGURE 4

Cash Flow of Rice Storage Project with 15% Inflation

End of Year	With Inflation			Without Inflation		
	Cash Flow	Debt Service	Net Cash Flow	Cash Flow	Debt Service	Net Cash Flow
		....	($'000)	....		
1	4.14	5.97	(1.83)	3.6	5.97	(2.37)
2	4.75	5.97	(1.22)	3.6	5.97	(2.37)
3	5.47	5.97	(0.50)	3.6	5.97	(2.37)
4	6.30	5.97	0.33	3.6	5.97	(2.37)
5	7.24	5.97	1.27	3.6	5.97	(2.37)
15	33.70	5.97	27.73	3.6	5.97	(2.37)
20	67.79	5.97	61.82	3.6	5.97	(2.37)

Relative size of investment and operating costs. Benefits from inflation are quite considerable in the case of projects involving large investment costs relative to operating costs as was the situation with the Rice Storage Project. In contrast, for projects involving large operating cost relative to initial investment, benefits from inflation are likely to be small because debt service would be a small part of annual operating cost. The next example illustrates this point.

A Service Cooperative running a tractor hire service borrows $10,000 from a bank to be repaid in three years at 15%, for the purchase of a tractor. The tractor is assumed to have zero residual value at the end of three years. Annual operating costs and revenues are estimated at $15,000 and $18,400 respectively. The loan of $10,000 would be repaid in 3 equal instalments.

Given the matching of inflation and borrowing rate at 15%, it may be argued that the project is profitable as shown below.

This example indicates that a project with high operating costs relative to investment cost has not much to gain from inflation.

FIGURE 5

Tractor Hire Project with Inflation

Price Index	End of Year	Investment Cost	Operating Cost	Revenue	Cash Flow	Debt Service	Net Cash Flow
			....	($'000)	....		
100	0	(10)	-	-	-	10	-
115	1	-	17.25	21.16	3.91	(4.38)	(0.47)
132	2	-	19.80	24.29	4.49	(4.38)	0.11
152	3	-	22.80	27.97	5.17	(4.38)	0.79

This example indicates that a project with high operating costs relative to investment cost has not much to gain from inflation.

Debt service deflated by inflation factor. The monetary gain from borrowing during periods of inflation can be demonstrated alternatively by **deflating the debt service by the inflation rate**, while keeping net cash flow at constant prices.

FIGURE 6

Rice Storage Project with 15% Inflation

Price Index	End of Year	Cash Flow	Debt Service (in Monetary Units)	Debt Service (deflated by Price Index)	Net Cash Flow (in Real Terms)
		($'000)			
100	0	-	-	-	-
115	1	6	5.97	5.19	0.81
132	2	6	5.97	4.52	1.48
152	3	6	5.97	3.93	2.07
175	4	6	5.97	3.41	2.59
201	5	6	5.97	2.97	3.03
936	15	6	5.97	0.64	5.36
1,883	20	6	5.97	0.32	5.68

As we can see, the results are the same as would occur if we use debt service in monetary terms while inflating the cash flow by the rate of inflation (see Figure 3). However, investors may find deflating the debt service difficult to understand. This is because **debt service figures are no longer in monetary units but in purchasing power units.** The use of a price index for deflating debt service means that the unit of measurement is changed from year to year. This is a confusing feature of this method. Furthermore, a **current purchasing power unit** is an abstract concept, not a physical object that can be exchanged. Although at any particular point of time a unit of current purchasing power may theoretically be exchanged for the same amount of goods and services covered by the price index (i.e. consumer price index or retail price index or other suitable index used to measure price changes) as a monetary unit (a dollar), and therefore can theoretically be exchanged for the monetary unit itself, this does not mean that it is the same as the monetary unit nor that it has the same attributes. Units of different currencies are not the same merely because a unit of one currency can be exchanged for a unit of another (Sandilands' Report on Inflation Accounting).

VI. PROBLEMS IN CASH FLOWS

Introduction

As already discussed, all the cash flows that arise directly or indirectly as a result of the project have to be identified. Some of the cost or benefit flows are easy to identify. For example, in Beta Ltd., costs of fixed assets - land, buildings, plant, machinery and equipment - as well as the revenues generated by the sale of vegetable oil and meals were obvious. However, there are problems in some cases which make it difficult to identify the relevant cash flows required for project appraisal. The purpose of this chapter is to deal with those items which generally cause difficulties.

Profits and Cash Flows

Under the accrual basis of accounting, profits will not necessarily match cash flows. Changes in profit can occur without any corresponding changes in cash flows depending on the yardsticks of measurement adopted for revenue recognition, identification of capital and revenue expenditure, method of depreciation and valuation of inventories. For example, funds from operations ($175,000) in Year 2 of Beta Ltd. (see Chapter V, Schedule 6) are not identical with net profit ($119,000) of that year. The difference is due to the depreciation charge of $56,000. Depreciation, as a measure of the fall in value of fixed assets (due to wear and tear and other factors) is merely a book entry and does not involve outflow of cash. Cash expenditures were incurred when the corresponding fixed assets were purchased. Similarly, write-offs or amortization of goodwill and other intangible fixed assets are charges against profit without involving an outflow of cash. A similar situation arises when assets like land and buildings are revalued to take account of inflation. The amount of surplus generated from a revaluation is added to the cost of the asset in question and a corresponding (book) entry is made by creating an account called Revaluation Reserve. Actual increase in cash occurs when the asset is disposed of, not when the entry is made.

The broader concept of **net operating cash flow** (NOCF), as opposed to **Funds from Operations** (FFO), can now be introduced, again using data from Beta Ltd. In its accounts, the amounts of revenue and expenses have not generated the same amounts of cash inflows or cash outflows in any of the 5 years. For example, no distinction is made between a cash sale and credit sale in the sales revenue. This means that the analyst who wants to determine the cash flow generated by the company's sales has to adjust the amount of reported sales to changes in Accounts Receivable so that cash flow from sales = sales for the period minus Accounts Receivable where Accounts Receivable denotes the difference between closing and opening balances of trade credits. This procedure can be repeated for all revenue and expense items involving cash transactions. For each item, take revenue or expense from the profit and loss statement and then make an adjustment by reference to corresponding closing and opening balances from the balance sheets. However, instead of making individual adjustments, a combined adjustment can be made to derive the net operating cash flow. For Year 2 of Beta Ltd. operations, the net operating cash flow will be $50,000 derived as follows.

	..$..	..$..
Sales revenue		1,418,000
Less cost of sales 1/	1,169,000	
Administrative & marketing costs 1/	20,000	
Short-term interest	4,000	
Taxes	Nil	
Change in working capital	175,000	1,368,000
Net Operating cash flow		= **50,000**

1/ By definition, non-cash charges like depreciation are excluded in these figures.

This differs from the sum of $175,000 representing funds from operations. The two figures can be reconciled as follows:

FFO = Profit + Depreciation
= Sales - Cost of Sales - Administrative & Marketing Costs - Interest charges - Taxes

or

FFO = NOCF + Working Capital - Long-term Interest Charges
i.e. 175,000 = 50,000 + 175,000 - 50,000 (Chapter V, Schedules 5 & 6).

Interest

Cash payments for interest are normally excluded from cash flows, because the interest element is taken into consideration while discounting the net streams of cash flows to obtain the IRR or NPV. To include also payments for interest would result in double counting. However, payments for short-term interest (supporting current operation of the enterprise) are included for cash flow calculation. It should be noted that long-term interest payments are excluded from the cash flow when discounting it to obtain return on equity.

Relevant Cash Flows

When considering relevant costs and benefits, account should be taken of the concept of **opportunity cost**. Some examples are given to explain this concept.

Wages. Cash outflow in respect of wages would be the opportunity cost to the firm in cases where the concerned workers like skilled persons are in short supply.

The way to calculate this cost is to add the cost of skilled workers and the loss of contribution due to these workers being drawn away from other operations in the enterprise. For example, annual wage for a new project is estimated at $45,000 for skilled workers who would otherwise be employed on other operations during a year. These workers would generate sales revenue of $120,000 in the other projects, causing variable expenses of $35,000 other than for skilled labour. By having these workers on the new project, the enterprise will lose a contribution of $40,000 as follows:

	. . $ . .
Lost sales from other projects	120,000
Less variable cost for skilled labour	45,000
Less other variable cost of production	35,000
Loss of contribution from other projects	**40,000**

Therefore the opportunity cost of these skilled workers to the new project is $40,000 (lost contribution) + $45,000 = $85,000. This will be entered as cash outflow in respect of wages and not $45,000.

Assets. Similar considerations apply where materials or equipment are in short supply. The opportunity cost will be the basis of cash flow and it will be calculated by adding to the cost of materials or equipment the lost contribution from other projects undertaken by the enterprise, which may have to be curtailed or completely shut down.

Sometimes material may be in stock but required by other projects. The opportunity cost is the cost of replacing the concerned stock. If, on the other hand, the enterprise had excess stock and would not have replaced it at any event, then the opportunity cost is the scrap value. This is given by the expected selling price less estimated cost of selling expenses.

Sunk costs. Sunk costs are those which have already been incurred before an **accept or reject decision** has been taken on the proposed project. Thus they have to be excluded from determination of cash flows relevant to a future project. Where new projects use capital assets left over from earlier projects, the cost to be considered in cash flow analysis is not the original cost but the prices which these assets will earn in the next best alternative. If they would be discarded as scrap, then it is the scrap value which is the opportunity cost and becomes the relevant value for the cash flow.

Apportioned overheads. Similarly, overheads already incurred for the general operation of the enterprise will not affect the decision on a new project, although for conventional historic accounting purposes, a portion is allocated to the new project. Allocated overheads will not therefore be included in the relevant cash flow. An example is given to illustrate these considerations.

Gamma Ltd. runs a furniture shop making office furniture out of rubber tree wood. As part of its expansion programme it has already spent $50,000 on research and development of rubber tree wood. The company plans to purchase a new plant costing $100,000. Costs of production for 5 years of new plant operation have been estimated variable costs at $5 per unit and fixed overheads (rent, rates of the company) apportioned at $20,000 per annum for this plant. Sales are forecast at 15,000 units per annum over the 5 years - all at a price of $7 each. Should the company purchase the new plant?

A cost analysis has been done on the basis of full cost and opportunity cost. The following results have been obtained.

Full Cost Basis
(5-year period)

Units produced and sold			**75,000**
		Unit Price	**Total Value**
Sales revenue		$7	525,000
Less variable costs		$5	375,000
Contribution		**$2**	**150,000**
Less Fixed Costs			
- Plant	$100,000		
- Research & development	$ 50,000		
- Other overheads	$ 20,000		170,000
Deficit			**20,000**

Opportunity Cost Basis
(5-year period)

		.... $ ....
Sales revenue		525,000
Less **Relevant Costs**		
- Variable costs	375,000	
- Fixed costs (plant cost)	100,000	475,000
Surplus		**50,000**

Although both presentations are correct, the full cost basis is not relevant for decision-making purposes. The sum of $50,000 spent for research and development has already been incurred before the decision to purchase new plant was taken. It is therefore a sunk cost, and irrelevant to the decision to purchase the plant. Similarly, fixed overheads will have to be borne by the company whether the new plant is bought or not. They were necessary for the existing operations of the company and cannot really be attributed to the new plant project. Based on full cost, the new plant is expected to result in a deficit of $20,000 while the opportunity cost basis produces a surplus of $50,000. It is therefore seen that cash flows based on opportunity costs give a proper basis for decision-making.

Residual values. Some assets can still remain and have a residual value at the end of project life. This residual value may be the value of land and scrap value of other assets like equipment or motor vehicles. The residual value must be estimated having regard to cost of demolition and dismantling before treating the difference as cash flows. Incremental working capital will be recovered in full at the end of project life subject to any losses on debtors and inventories.

After-tax flows. Profits after tax are only relevant for investment appraisal. This is because, given the assumption of maximization of shareholder wealth, cash flows available to the shareholder would be the guiding factor in the appraisal of an investment project. Tax laws normally provide reliefs by way of accelerated depreciation, stock relief and treatment of business related expenses including interest payments as allowable expenses. Depreciation and stock reliefs are primarily intended to prevent a tax based on historic cost profits from eroding the company's capital in time of inflation. Under inflation the tax relief on interest is, in part, in effect a relief on the repayment of capital because the interest rate is generally higher in order to make some allowance for the loss of value of the capital through inflation.

The tax computation is, however, based on the company's net profits rather than on its net cash flow. Net profits of a company have to be established before its tax liability can be assessed and included in the cash flows. Taking the example of Beta Ltd., and assuming a company tax rate of 50%, the company's tax is calculated as shown below.

Computation of Company Tax

Year	Revenues	Operating Costs without Depreciation	Depreciation	Net Profit	Tax at 50%
		....................($'000)....................			
1989	-	-	-	-	-
1990	1,418	1,243	56	119	60
1991	1,575	1,362	56	157	79
1992	1,575	1,341	56	178	89
1993	1,575	1,324	56	195	98
1994	1,575	1,308	56	211	105

Assuming a one year lag in tax payment, the company's net cash flows can be established by incorporating the tax flows in Beta Ltd., Schedule 8 (Chapter V).

Year	Investment Cost	Cash Flows from Operations	Tax	Net Cash Flow
	....	($'000)	....	
1988	800	-	-	(800)
1989	207	225	-	18
1990	18	263	60	185
1991	-	267	79	188
1992	-	267	89	178
1993	(375)	267	98	544
1994	-	-	105	(105)

Financial Rate of Return (after tax) = 5%

Inflation Factor in Cash Flows

Up to this stage in the analysis of cash flows, we have assumed price levels to be stable and estimated project costs and revenues in **constant prices** for IRR calculations. However, **price changes** are a fact of life which should be reflected in the future cash flows of a project. The discount rate used to reduce the cash flows to present value may either be market or real rate of interest. The general relationship can be expressed as (1 + real rate of interest) x (1 + inflation rate) = (1 + market rate of interest). Correct project evaluation can be made using either of the following methods. First NPV may be calculated by estimating future money cash flows and discounting at the expected market rate of interest. Alternatively, the evaluation can be carried out by estimating future cash flows in real terms and discounting at real rate of interest. These two approaches are illustrated by the following example.

Suppose that a project, at Year 3, produces a cash flow of $5,000. The present value of this cash flow can be found by either of the above methods, given the market rate of interest at 12% and the annual rate of inflation at 8%. We can find the real rate of interest to be 3.7% computed as follows:

$$(1 + \text{real rate of interest}) \times (1 + 0.08) = (1 + 0.12)$$

$$\text{Real rate of interest} = \frac{1 + 0.12}{1 + 0.08} - 1$$

$$= 0.037 \text{ or } 3.7\%$$

The present value of the cash flow can now be found through the two approaches.

1. **Money cash flow discounted by market rate of interest**

$$\text{Present value} = \frac{\$5{,}000}{\left(\frac{1}{(1 + 0.12)}\right)3} = \$3{,}559$$

2. **Real cash flows discounted by real rate of interest**

$$\text{Present value} = \frac{\$5{,}000}{\left(\frac{1}{(1 + 0.08)}\right)3 \times \left(\frac{1}{(1 + 0.037)}\right)3} = \$3{,}559$$

The present values are obviously the same in both cases. In the first approach, the cash flow has been discounted by 12% while in the second, the cash flow has been first deflated by inflation rate of 8% to obtain real cash flows and then discounted by 3.7% which is the real rate of interest.

It is worth noting that the project cash flows are deflated by the general rate of inflation rather than a specific rate that applies to a particular cash flow. For example, suppose that although the general rate of inflation as reflected in the consumer price index is 10%, the rate of inflation of vegetable oil is 15%. The actual money cash flow received from oil sales would be deflated by 10% and not by 15%. This is because the money cash flows of a project are expressed in terms of their ability to purchase a collection of goods and services and not their capacity to purchase any one particular goods or service.

VII. PRICING AND PROFIT STRATEGIES

Pricing Strategies

There are basically two ways to determine the price of a product or service - through an assessment of the market or an analysis of the costs of production. Obviously, the market is the final determinant; however, there will be cases of price fixing using production costs. There are several methods of estimating prices on this basis: **marginal cost pricing, full cost pricing, rate of return pricing, conversion cost pricing** and **transfer pricing.**

Ascertainment of Marginal Cost (MC)

Marginal cost is the variable cost of one unit of a product or a service, i.e. a cost which would be avoided if the unit was not produced or provided. The marginal cost can be calculated as follows:

Marginal Cost of Product X

Additional Direct Costs for One Unit	. . $ . .
Materials	0.70
Wages	0.10
Expenses	0.25
Prime cost	**1.05**
Variable Overhead Cost per One Unit	
Production	0.15
Marketing and distribution	0.20
Administration	0.05
Overhead cost	0.40
Marginal cost	**1.85**

Product and Market Profitability

Marginal costing is particularly effective in providing information which would help management in selecting products, markets, sales areas, and classes of customers. A simple illustration is given below to illustrate product profitability. China Forest Ltd. produces two types of plywood - X and Y. Both products worked on are finished and sold. The following is the profit statement for the year 31.12.90.

	Total	Product X	Product Y
	....	($)	....
Sales (units)	**1,500**	**800**	**700**
Sales revenue	23,000	16,000	7,000
Direct material	11,500	8,000	3,500
Direct labour	5,400	4,000	1,400
Prime cost	**16,900**	**12,000**	**4,900**
Production overheads	3,100 1/	2,000	1,100
Production cost	**20,000**	**14,000**	**6,000**
Marketing, distribution and administration cost	2,200 2/	1,000	1,200
Total Cost	**22,200**	**15,000**	**7,200**
Profit/loss	**800**	**1,000**	**(200)**

1/ Variable $1,700 ($900 for X and $800 for Y) and fixed $1,400 apportioned - X $1,100 and Y $300.

2/ Variable $500 ($300 for X and $200 for Y) and fixed $1,700 apportioned - X $700 and Y $1,000.

This would suggest that product Y be discontinued. But this ignores the fact that whether or not Y is manufactured, certain fixed costs such as factory rent, rates, machine depreciation and administration salaries will have to be met. Marginal costing presentation would clarify the position enabling a correct decision to be taken as shown below.

	Total	Product X	Product Y
	....	($)	....
Sales revenue	23,000	16,000	7,000
Less **variable costs**			
Direct material	11,500	8,000	3,500
Direct labour	5,400	4,000	1,400
Variable production overheads	1,700	900	800
Variable marketing, distribution & administration overheads	500	300	200
Total variable costs	**19,100**	**13,200**	**5,900**
Contribution	3,900	2,800	1,100
Less **fixed overheads**			
Production overheads	1,400	-	-
Marketing, distribution and administration overheads	1,700	-	-
Total Fixed Overheads	**3,100**		
Profit	**800**		

From the above, it is clear that product Y brings in a contribution of $1,100. This would be lost if its production is discontinued; but the fixed cost of $3,100 would still be incurred. So an overall loss of $300 would be incurred if Y were to be discontinued instead of a total profit of $800 when both are produced. Unless fixed costs can be reduced by more than $1,100 if product Y is discontinued, production of Y should be continued.

Combined effect. Marginal analysis facilitates tracing the combined effect of price and cost factors on profit. To illustrate this, let us take an income statement of an enterprise for two years.

	Year 1 ($)	Year 2 ($)
Sales	200,000	400,000
Marginal cost of sales	100,000	150,000
Contribution	**100,000**	**250,000**

The changes from Year 1 to Year 2 are attributed to improved production methods, a rise in selling price by 25% and an increase in the quantity sold. The question is to trace the effect of individual factor which caused an increase of contribution by $150,000.

Change due to volume. Sales price increased by 25% or ¼, therefore:

Sales of Year 2 at Year 1 price = $400,000 x $\frac{4}{5}$ = $320,000

Sales of Year 1 at Year 1 price = $200,000

Change due to volume = $120,000

Percentage change in volume = ($\frac{120}{200}$ x 100%) = 60%

Sales increase = $120,000

Marginal cost = <u>$60,000</u> ((60% x 100,000))

Contribution change due to volume = **$60,000**

Change due to selling price:

Sales in Year 2 at Year 1 price = $320,000

Sales in Year 2 at Year 2 price = <u>$400,000</u>

Contribution change due to price = **$80,000**

Reduction in cost:

Change in sales volume = ($\frac{120,000}{200,000}$ x 100%) = 60%

Marginal cost in Year 1 = $100,000

Marginal cost in Year 2 should be = $100,000 + ($\frac{60}{100}$ x 100,000)

= 160,000

But marginal cost in Year 2 was = $150,000

Reduction in cost = $10,000

Thus the change in contribution of $150,000 is due to:

Volume change of	$60,000
Price change of	$80,000
Cost change of	<u>$10,000</u>
	$150,000

Limiting factor. Every company has one or more limiting or key factors, that is, a critical input for a business which at a particular point in time or over a period will limit the scale of its operations. It is frequently sales potential but it may also consist of a certain class or type of raw material or plant, skilled labour, floor space or liquid resources. Where such constraints operate, profitability should be determined by the level of contribution per limiting factor. Linear programming provides an efficient mathematical search procedure for selecting the optimum plan where there are a number of limiting factors and interacting variables.

Marginal Cost Pricing

In times of recession or low demand for a product, marginal costing is often used as a basis for determining prices. For example, during a slump, it may be desirable to accept orders below total cost on the grounds that it they could cover marginal costs, any contribution towards fixed expenses would at least reduce losses and keep together the facilities and employees in the hope of better times to come.

The marginal costing techniques can thus help management in fixing prices in such special circumstances as:

(a) a trade depression in the industry;

(b) spare capacity in the factory;

(c) a seasonal fluctuation in demand;

(d) when a special contract is being sought;

(e) where alternative levels of activity are being considered.

Absorption, Full Cost or Cost Plus Pricing

The usual procedure in conventional full cost pricing (as opposed to marginal cost pricing) is for the company to calculate the cost of producing a unit of each product at the normal capacity level of its existing plant and then add to this unit cost what it regards as the most satisfactory profit margin in relation to its competitors' prices. This procedure is illustrated below.

Suggested Selling Price

	Product A ($)	Product B ($)
Direct material	5	10
Direct labour	4	2
Direct expense	1	-
Prime cost	**10**	**12**
Production overhead:		
- Variable	5	5
- Fixed	5	10
Cost of production	**20**	**27**
Marketing and distribution cost:		
- Variable	2	1
- Fixed	1	2
Administration overhead:		
- Fixed	1	1
Total Cost	**24**	**31**
Profit margin	10%	20%
Selling price	26.40	37.20
Marginal Cost (i.e. total variable costs)	**17**	**18**

Rate of Return Pricing

This method of setting prices is usually employed by nationalized industries, and is targeted at obtaining a planned rate of return on capital employed. To translate this rate of return into a percentage mark-up on costs (i.e. to find profit margin), it is necessary to estimate a 'normal' rate of production. Total costs of a year's normal production are then examined and these are taken a the total annual cost in the computation. The basic formulae for the rate of return pricing is:

$$\frac{\text{Capital employed}}{\text{Total annual cost}} \times \frac{\text{Profit}}{\text{Capital employed}} = \frac{\text{Profit}}{\text{Total annual cost}}$$

Therefore:

$$\frac{\text{Percentage}}{\text{Mark-up on cost}} = \frac{\text{Capital employed}}{\text{Total annual cost}} \times \text{Planned rate of return}$$

Illustration of Rate of Return Pricing

Zambia Mines Ltd. makes the following estimates:

Variable costs	\$20 per unit
Fixed costs	\$50,000 per year
Normal production	24,000 units

Normal capital employed:

Variable	\$5 per unit
Fixed (\$)	\$750,000
Desired rate of return	12%

The percentage mark-up on cost is calculated as follows:

$$\frac{25{,}000 \times 5 + 750{,}000}{25{,}000 \times 5 + 50{,}000} \times 12\% = 19\%$$

The selling price per unit which will produce the desired rate of return of 12% is calculated as follows:

	(\$)
Variable cost per unit	20
Fixed cost per unit (50,000-25,000)	2
Total Cost	**22**
Mark up cost (19%)	4
Selling price	**26**

Conversion Cost Pricing

This method of price-setting directs the sales effort to those product which require the least amount of labour costs and overheads. Labour costs and overheads are the components of conversion cost, i.e. costs which convert materials into finished products.

Suppose a company produces two products each selling for \$10 per unit. The cost of production for each is \$9 giving a profit of \$1. This would indicate that from a profit point of view, the company would be indifferent between producing one or the other. A decision on this basis may prove sub-optimal if the cost structure of the two products varies as the following data shows:

	Product A ($)	Product B ($)
Materials	6	3
Labour	2	4
Overheads	1	2
Cost of Production	**9**	**9**
Selling Price	10	10
Profit	**1**	**1**

The cost breakdown shows that product A requires only half the labour required by product B, overheads are similarly one half. If it were possible to transfer all efforts to A, a greater number of units could be produced and sold with the same selling price but with much larger profits.

Transfer Pricing

A company may operate through several autonomous divisions, each divisional manager being held responsible for the profit performance of his division. Problems arise where Division A supplies products to Division B and the manager of Division B has the power to purchase these products from external suppliers. If the latter course is followed because outside prices are lower, the profits of Division B will rise and the profits of Division A will fall due to the reduced volume of output. The question arises as to what price should Division A charge for supplying products to Division B so that the correct action in terms of achieving the best overall profit for the company and yet not unduly prejudice the profit performance of each divisional manager. Problems arise when the products are of such a nature that there is no external market price to use as a yardstick. Different and less satisfactory methods (cost-based prices) will be adopted in the absence of reliable market prices.

Profit Strategies

Break-even analysis. In project planning, it will be helpful to examine the financial consequence of pricing policies and cost changes on project sales and profitability. Break-even (B/E) analysis assist in developing this relationship. Break-even analysis is aimed at establishing the relationship between fixed costs, variable costs, sales revenue and profit. The analysis normally employs a B/E chart which indicates the approximate profit or loss at different level of sales volume. The point at which sales revenue and total costs are exactly equal is termed the B/E point - point **X** in Chart 1. The B/E point may also be calculated by formulae as follows:

$$\text{B/E in sales unit} = \frac{\text{Total fixed cost}}{\text{Contribution per unit}}$$

$$\text{B/E in sales value} = \frac{\text{Total fixed cost} \times \text{Total sales}}{\text{Total contribution}}$$

or

$$\frac{\text{Total fixed cost}}{\dfrac{\text{Total contribution (C)}}{\text{Total sales (S)}}}$$

CHART 1

BREAK-EVEN CHART

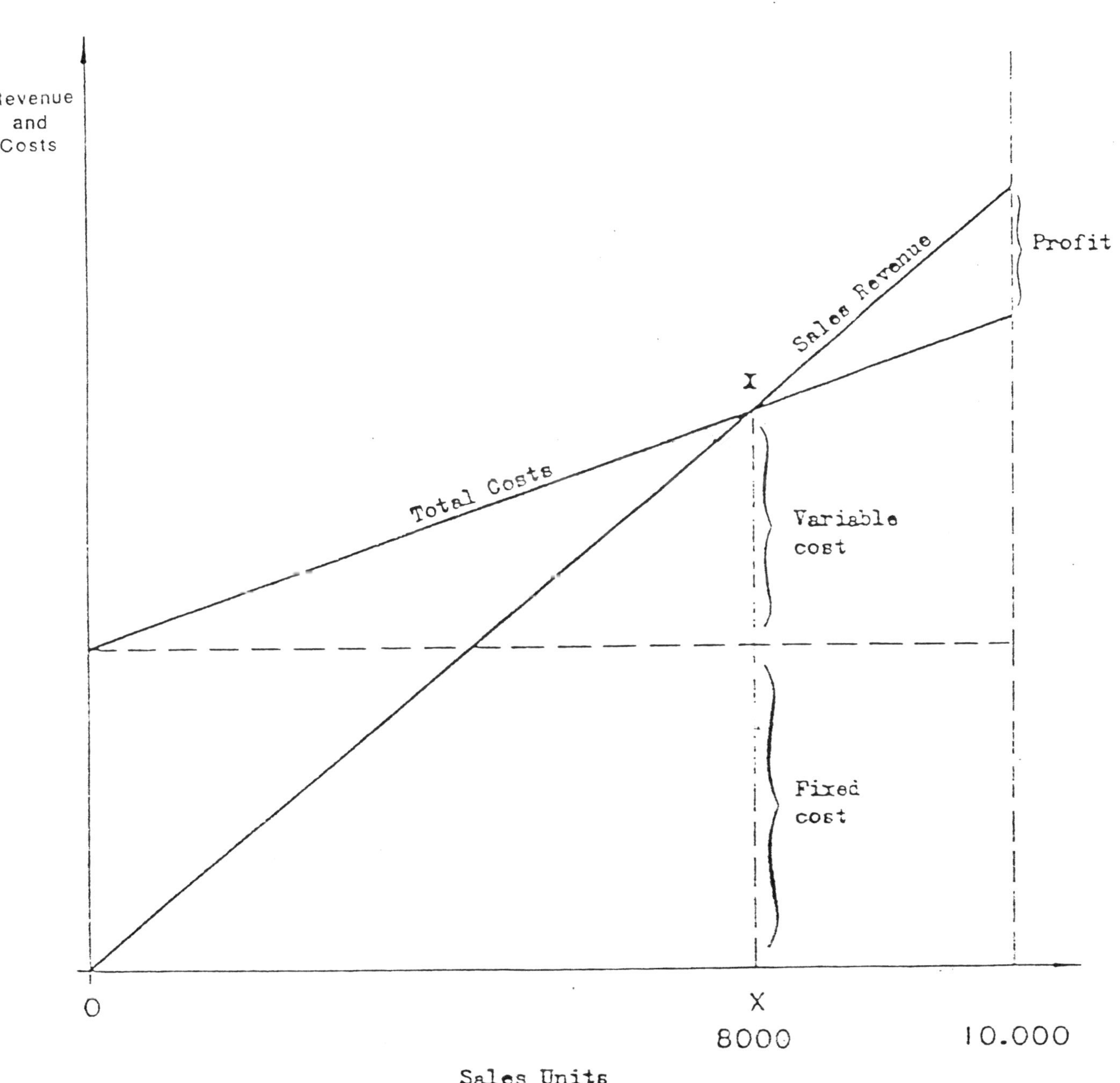

A Simple Illustration

The current costs and sales of Zambia Canned Fruits Ltd. at 100% of normal capacity are:

	($'000)
Annual sales revenue (unit price $10)	100
Variable costs ($5 per unit)	50
Fixed costs	40

This information can be presented a follows:

	($'000)
Sales revenue	100
Variable costs	50
Contribution	**50**
Fixed costs	40
Profit	**10**

The excess of sales value over variable costs is the contribution ($50,000) which goes to meet fixed costs ($40,000) and generates profit ($10,000); and where the contribution is exactly equal to total fixed costs, then the enterprise is said to break even. The break-even values are computed using the above formulae as follows:

$$\text{B/E in sales unit} = \frac{40{,}000}{5} = 8{,}000 \text{ units}$$

$$\text{B/E in sales value} = \frac{40{,}000 \text{ x } 100{,}000}{50{,}000} = \$80{,}000$$

$$\text{B/E capacity in \%} = \frac{\text{B/E Sales}}{\text{Actual Sales}} = \frac{80{,}000}{100{,}000} \text{ x } 100\% = 80\%$$

The contribution to sales (C/S) ration (or profit volume ratio or simply PV ratio) is useful to calculate break-even capacity and sales and also the effect of changes in sales volume on profit. C/S ratio for the Zambian Company is:

$$\frac{50{,}000}{100{,}000} \text{ or } 50\%$$

An alternative way or presenting the same information is by a profit volume chart (see Chart 2). This chart shows that with 'nil' sales the company has a loss equivalent to its fixed cost. When it generates sales and incurs variable costs, it produces a contribution. When the latter equals fixed cost, B/E is reached - at point B in Chart 2.

CHART 2

PROFIT-VOLUME CHART

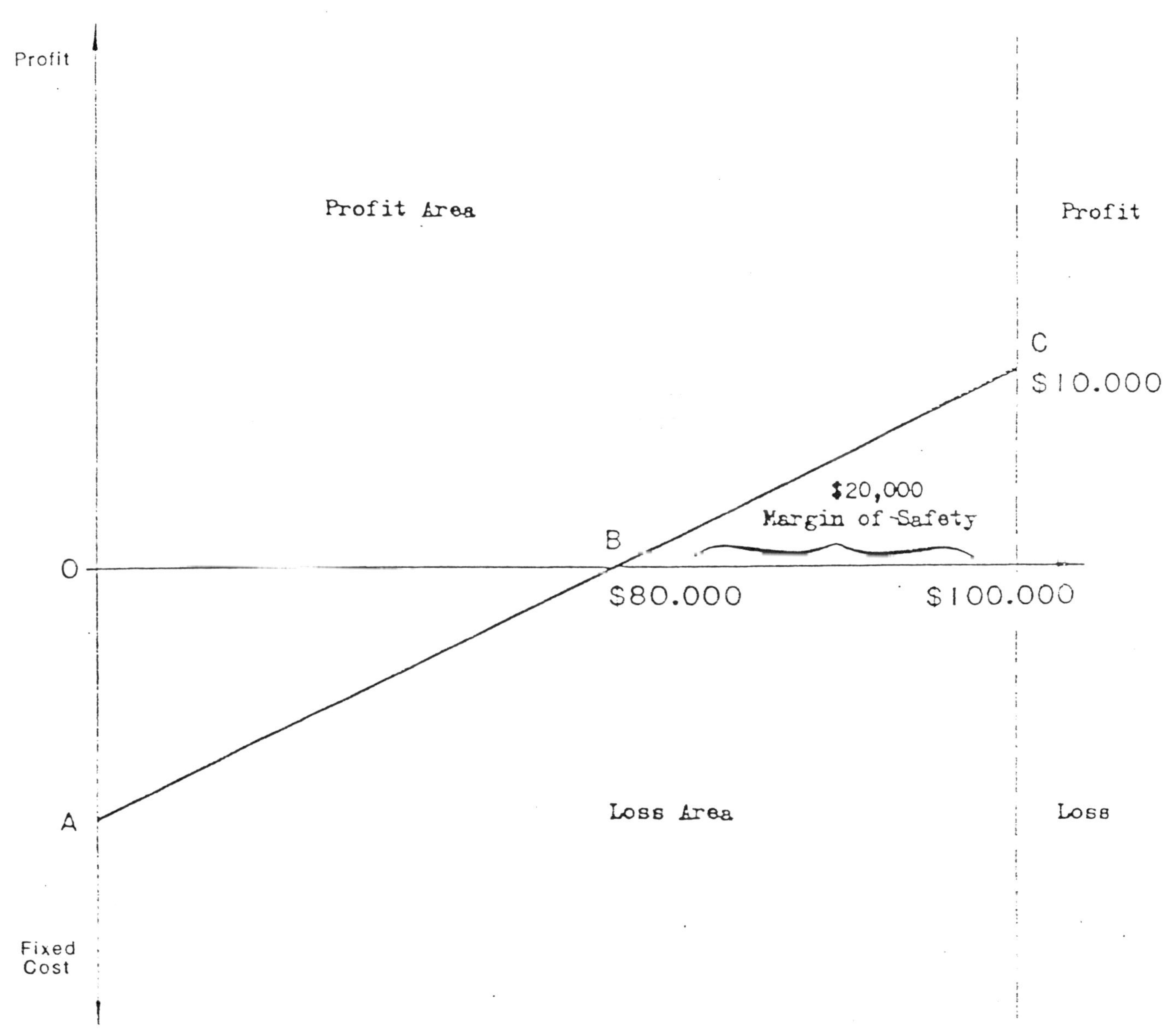

As shown above, the B/E chart is a useful device for presenting a simplified picture of cost-volume-profit relationships. However, the chart is based on many restrictive assumptions which are:

(a) Costs are either fixed or variable, or at least they can be so identified and classified for purposes of B/E analysis. Where cost figure contains fixed and variable elements (known as semi-variable or semi-fixed cost), for example, maintenance cost, segregation is possible through use of methods such as Range (high and low points), Scattergraph, or least squares analysis.

(b) Fixed costs and variable costs are correctly separated over the whole range of output.

(c) Selling price per unit is constant regardless of the level of output or the demand curve for the product is known.

(d) There is one product or if there is more than one product, a constant sales mix exists.

(e) Production and sales are equal and there is no change in finished goods stock.

(f) Volume is the only factor affecting costs and all other factors remain constant.

It must also be emphasized that the B/E chart does not take into account capital employed, thereby limiting its usefulness as a measure of profitability.

These limitations do not mean that B/E analysis is completely useless. The greatest value comes from the analysis of the underlying relationship of volume, costs and profits and not from the location of the B/E point. The point itself is of limited significance. The proximity of the B/E point to actual sales may influence management's attitudes towards return and risks and towards need for cost reduction efforts. One measure of proximity of the B/E point is margin of safety (MS) which is the difference between actual sales and B/E sales or in percentage terms:

$$\frac{\text{Actual Sales} - \text{B/E Sales}}{\text{Actual Sales}} \times 100\%$$

The margin of safety is important in times of recession characterized by falling sales. The greater the margin of safety, the further sales can fall before the B/E is reached. Using the above illustration, the margin of safety on sales volume is $20,000 (100,000 minus 80,000),

or 20% $\left(\frac{100{,}000-80{,}000}{100{,}000} \times 100\%\right)$

The relationship between margin of safety, contribution, sales and profit can be shown by means of an equation:

$$\text{Profit (\%)} = \text{MS} \times \text{C/S}$$

This can be proved as follows:

Profit as a percentage of sales in the above illustration is 10% which is equal to:

$$\text{MS} \times \frac{C}{S} = 20\% \times \frac{50}{100} = 10\%$$

Improving Profitability

Profits of an enterprise can be improved by raising selling prices, increasing sales volume, reducing variable and fixed costs, changing sales mix or a combination of these. Break-even analysis helps the analyst to identify the key variables and examine the effects of changes on profits. The likely changes are illustrated by simple examples.

Price-volume relationship. The effects of a change in price as compared to volume on profits are illustrated by an example. Three companies A, B and C have the cost-volume profit profiles as follows:

	A	**B**	**C**
Annual Sales Volume (units)	120,000	60,000	6,000
	....	($)	
Sales revenue	960	720	120
Variable costs	560	360	40
Contribution	**400**	**360**	**80**
Fixed cost	300	260	50
Profit	**100**	**100**	**30**
B/E sales volume (units)	90,000	43,333	3,750
B/E sales value ($)	720,000	520,000	75,000
C/S ration	41.7%	50%	66.7%

The relationship between the C/S ratio for the three companies and the relationship between price and volume increases is shown below.

Company	C/S Ratio	% Price Increase which Equates to 10% Volume Increase
A	41.7%	4.2%
B	50.0%	5.0%
C	66.7%	6.7%

It is clear that the lower the C/S ratio, the greater the advantage of price increase over volume increase.

The following table compares a 10% increase in price with a 10% increase in volume for the company B. A 10% price increase gives $36,000 more profit than 10% volume increase. The price increase reduces B/E sales while the volume increase does not affect B/E sales. For the company B, a price increase on 5% equates to a 10% volume increase.

Company B
Comparison of 10% Price Increase with 10% Volume Increase

	Current Position	Increase	Position with Increase
	($)		
10% Volume Increase			
Sales revenue	720	72	792
Variable costs	360	36	396
Contribution	**360**	**36**	**396**
Fixed costs	260	-	260
Profit	**100**	**36**	**136**
C/S	50%		54.5%
Break-even			
Volume	43,333		43,333
Sales ($)	520,000		520,000
10% Price Increase			
Sales revenue	720	72	792
Variable costs	360	-	360
Contribution	**360**	**72**	**432**
Fixed costs	260	-	260
Profit	**100**	**72**	**172**
C/S	50%		50%
Break-even			
Volume	43,333		36,111
Sales ($)	520,000		476,667

Impact of cost increases. The same illustration can be used to illustrate the effect of cost increases in fixed and variable costs.

Company B
Effect of 10% Price Increase in Costs

	Current Position	**Increase**	**Position with Increase**
	....	($)	....
Sales revenue	720	-	720
Variable costs	360	36	396
Contribution	**360**	(36)	**324**
Fixed costs	260	26	286
Profit	**100**	**62**	**38**
C/S	50%		45%

Additional sales required to recover a 10% increase in costs would be $137,778 or 19.1% of current sales.

This is computed as follows:

$$\frac{\text{Additional profit required}}{\text{C/S}}$$

$$= \frac{\$62{,}000}{45} \times 100 \quad = \$137{,}778$$

A comparison with the other companies, A and C shows that the lower the C/S ratio the greater the impact of cost increases on company profitability as shown below.

Company	**C/S Ratio**	**To Recover 10% Increase in Costs**	
		Volume Increase	**or Price Increase**
A	41.7%	25.0%	9.0%
B	50.0%	19.1%	8.6%
C	66.7%	11.8%	7.5%

Proportion of fixed costs to total costs. Costs tend to become increasingly fixed in relation to short-run changes in activity. For example, labour costs may assume the characteristics of fixed costs. How does a high proportion of fixed costs in total costs affect decision making? Let us illustrate this with examples of Bangladesh Marketing (BM) which has a low fixed cost as opposed to Malaysian Electricity Company (MEC) which has a high fixed cost.

	100% Capacity	
	BM ($)	**MEC ($)**
Annual sales revenue	500	500
Variable costs	400	100
Contribution	**100**	**400**
Fixed costs	50	350
Profit	**50**	**50**
Ratio of fixed cost to total cost	11%	78%
C/S	20%	80%
B/E sales	$250,000	$437,500
B/E capacity	50%	87%

It will be observed that MEC has a high ratio of fixed cost to total cost, has a high C/S ratio and higher B/E sales and capacity utilization. What would be the effect of a 10% increase in capacity utilization?

Assuming all other factors remaining constant, an increase of capacity would have the following results:

	BM	**MEC**
Increase in sales	$50,000	$50,000
C/S ratio	20%	80%
Increase in contribution	$10,000	$40,000
Increase in fixed costs	nil	nil
Increase in profit	$10,000	$40,000

The above results show that the higher the ratio of fixed costs to total costs, the higher will be the C/S ratio and break-even sales.

The higher the C/S ratio, the more volatile will profits be in response to changes in capacity utilization. For example, when capacity utilization decreases to 40% of current capacity, losses would be $10,000 (BM) and $190,000 (MEC), changes of 120% and 480% respectively as derived below.

	BM	MEC
	.... ($'000)	
Annual sales revenue	200	200
Variable costs	160	40
Contribution	**40**	**160**
Fixed costs	50	350
Loss	**10**	**190**

Impact of a change in sales mix. An increase in profits can be achieved by concentrating on products with a relatively high contribution to sales ratio as seen in the Kenya Foods Ltd.

Kenya Foods Ltd. has a sales budget as follows:

Product	Sales	Variable Costs	Contribution
	....	($'000)	....
A	200	100	100
B	500	200	300
C	250	50	200
D	50	50	-
Total	**1,000**	**400**	**600**

C/S ratio = 60%

The current mix is: A 20%, B 50%, C 25% and D 5%.

You are required to identify the product which gives the maximum profit and comment on the results when sales mix is changed to: A 85%, B 5%, C 5% and D 5%.

Product C is the most profitable as can be seen below:

Product	C/S
A	50%
B	60%
C	80%
D	0%

When the sales mix is changed, as proposed, the new contribution will be $495,000 and C/S 49.5%, which are lower than the original results. The reason for this is that the sales of products having a higher C/S have been reduced in favour of product A which has a lower C/S ratio.

Product	Sales	Variable Costs	Contribution
		.... ($'000)	
A	850	425	425
B	50	20	30
C	50	10	40
D	50	50	0
Total	**1,000**	**505**	**495**

Assuming a fixed cost of $200,000, the B/E point for the company before and after changes in sales mix can be computed.

B/E in sales value before changes = $\frac{200,000}{\frac{600,000}{1,000,000}}$ = $333,333

B/E in sales value after changes = $\frac{200,000}{\frac{495,000}{1,000.000}}$ = $404,040

VIII. RISK ANALYSIS

Project investment can be characterized by reference to certainty, risks and uncertainty. Certainty, however, seldom exists for expected or future returns on investments. Uncertainty refers to an event, such as technological breakthrough resulting in obsolescence, that is expected to take place although the probability and timing of its occurrence cannot be forecasted during the life of the project. Although uncertainties can be treated in a descriptive manner, the data are insufficient to incorporate them into the analysis in a quantitative manner. Risk refers to a situation in which a probability distribution of future returns can be established for the project. The riskiness of project can be defined as the probable **variability of its future returns**. For a variable such as sales from which future returns are derived, sales risk can be defined as the variability of expected sales. In practice, during project analysis, there are usually several variables for which doubts exist as to their best estimates. The purpose of risk analysis is to isolate the risks and to provide a means by which the various project outcomes can be reduced to a format from which a decision can be made. The final result of risk analysis is a judgement regarding the possible range of future returns, as well as the likelihood of each value within this range.

While this paper is confined to determining the riskiness of a particular project, there are instances which require the examination of overall risk, which is also referred to as **portfolio risk.** The company's portfolio risk can be analysed by considering the relationship between the proposed project and the existing investments and operations of the company. While this type of analysis is beyond the scope of this paper, it should be noted that a project which may have been rejected when appraised alone, could be acceptable if evaluated alongside present investments because the overall risk would be reduced. Similarly, a project accepted when considered in isolation could be rejected because it would increase overall risk.

Methods for Quantifying Risk

There are three common methods for quantifying risk:

- **Probability distribution**;
- **Simulation**; and
- **Sensitivity analysis.**

Probability Distribution

Risk is associated with variability of returns - the more variable the expected future returns, the riskier the investment. However, it is useful to define risk more precisely.

Any investment decision - or any kind of business decision - implies a forecast of future events. Earlier illustrations have shown the expected annual cash flow as a point estimate - single figure, frequently called the 'most likely' or 'best' estimate. How confident is the analyst of his forecasted cash flows? The degree of uncertainty can be defined and measured in terms of the analyst's probability distribution - probability estimates associated with each possible outcome. In its simplest form, a probability distribution could consist of an optimistic, pessimistic and a most likely estimate or alternatively, high, low and a 'best guess' estimates. One could expect the optimistic or high estimate to be realised if the marketing prospects of the company are booming, pessimistic or low estimates to prevail under conditions of market recession, and most likely or best guess estimate to prevail during normal conditions. But how likely are these events - boom, recession and normal conditions? These events have to be given estimates of probability so that both a weighted average cash flow estimate can be developed. This point is best illustrated by an example.

A cotton mill has two investment proposals, X and Y, with the following cash flows:

Pay-off matrix for X and Y

State of the Market	**Annual Cash Flows**	
	Project X	**Project Y**
	. . . . (S)	. . . .
Boom (high)	1,200	2,000
Normal (most likely)	1,000	1,000
Recession (low)	800	0

The probability of a boom is 0.2, the probability of normal conditions is 0.6, and the probability of a recession is 0.2.

Given the annual cash flows under the three possible states of the future market and their probability of occurring, weighted average projected cash flows can be calculated by multiplying each cash flow by its probability of occurrence. This is done below:

Calculation of expected values

Appraisal	Cash Flow	Probability	Weighted Average or Expected Value
Project X			
Boom	1,200	0.2	240
Normal	1,000	0.6	600
Recession	800	0.2	160
			expected value 1,000
Project Y			
Boom	2,000	0.2	400
Normal	1,000	0.6	600
Recession	0	0.2	-
			expected value 1,000

The last column of this table is added to obtain a weighted average of the outcome for each alternative under the three events - boom, normal and recession. This weighted average is defined as the **expected value** of the cash flow from the investment, i.e. the 'mean' outcome of a range of possible values which are assigned to a particular alternative course of action.

Risk inherent in business decision-making can be displayed in graphical form to obtain a clear picture of the variability of actual outcomes. For example, assume two mutually exclusive investments **R** and **S** which have the same net cash flow of $4,000. However, investment **R** has a probability of cash inflows in the range of $3,000-$9,000 while investment **S** has a probability in the range of $0-$12,000. Looking at the net cash flow (which is equal in both cases), one is tempted to rank both investments equally. But a look at the graph below shows that the variability of cash flows for **S** is greater than that of **R** and therefore riskier.

PROBABILITY OF CASH FLOW

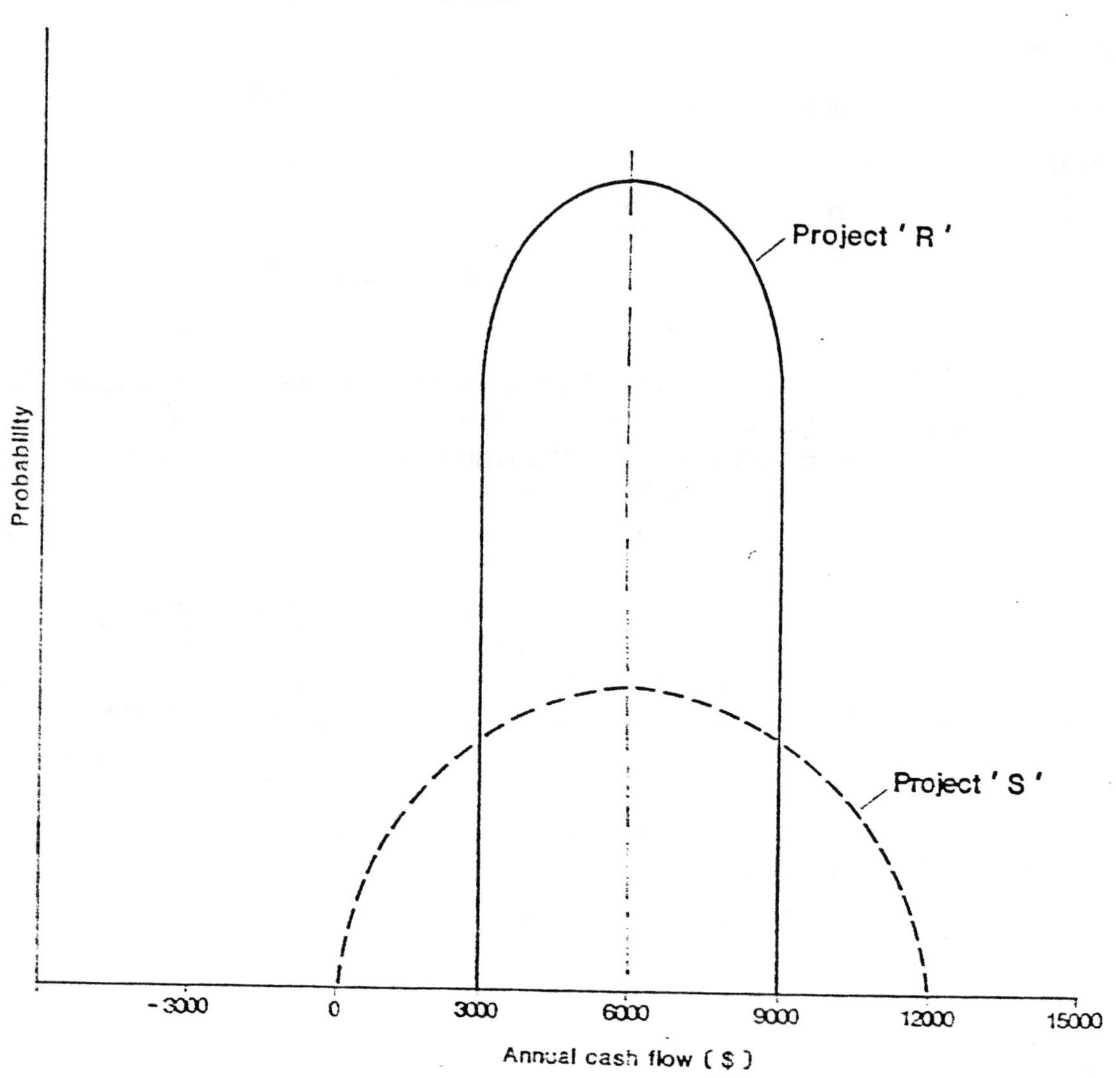

The variability of a distribution is normally measured by **standard deviation.** The earlier example is used to calculate the standard deviation as presented below.

Calculation of Standard Deviation

Probability	Net Cash Flow	Expected Value (1 x 2)	Deviation	Standard Deviation	Variance (1 x 5)
1	2	3	4	5	6
		....	($)	....	
Project X					
0.2	1,200	240	200	40,000	8,000
0.6	1,000	600	0	0	0
0.2	800	160	(200)	40,000	8,000
	expected value =	**1,000**		**variance =**	**16,000**
				Standard deviation =	**$126.49**
Project Y					
0.2	2,000	400	1,000	1,000,000	200,000
0.6	1,000	600	0	0	-
0.2	0	0	(1,000)	1,000,000	200,000
	expected value =	**1,000**		**variance =**	**400,000**
				Standard deviation =	**$632.45**

Columns 1, 2 & 3 are taken directly from the previous example. In column 4, the expected value from each possible outcome (i.e. net cash flow) is subtracted to derive the deviations about the expected value. In column 5, deviations are squared. In column 6, squared deviation is multiplied by the associated probability and the products are summed to obtain the variance of the probability distribution.

The standard deviation is then calculated by taking the square root of the variance, which gives $126.49 for Project X and $632.45 for Project Y. Project Y is much riskier as its standard deviation is much larger than the standard deviation for Project X.

Certain problems may arise in the use of standard deviation as a measure of risk. Taking an example which shows the probability distributions for investments M and N, assume investment M has an expected return of $2,000 and a standard deviation of $600, and investment N also with the same standard deviation of $600 has a higher expected return of $8,000. Investment M has more risk per dollar of return than N. On this basis, it is reasonable to assign a higher degree of risk to investment M than to investment N, even though both have identical standard deviations. The standard procedure for handling this problem is to divide the standard deviation by the expected value of net cash flows to obtain the coefficient of variation. For investment M, $600 (standard deviation) is divided by $2,000 (expected value), obtaining 0.30 as the coefficient of variation. Similarly, investment N has a coefficient of variation of 0.075. Since N has a much lower coefficient of variation, it has less risk per unit of return than investment M.

Simulation

When cash flows are correlated over time, the standard deviation calculation will not give a correct calculation of the variation of the project's NPV. Also, in practice, it may be necessary to produce separate probabilities for alternative sales revenue, different items of costs, and different possible life spans. These problems can be overcome by using Monte Carlo simulation analysis. Simulation trials are tried out by computer, which allows the consideration of many variables and their probability distributions, and which provides a large amount of information on which to make an investment decision. Based on the inputs, simulation analysis can provide probability distributions for cash flows, IRR and NPV. Certain conclusions can be arrived at such as: 'the likelihood is 80% that the IRR will equal 12%' or there is a 10% chance that the project will result in a loss of $200,000.

Sensitivity Analysis

The purpose of sensitivity analysis is to identify the variables to which the NPV is most sensitive, and it is normally used to measure the extent to which these variables can change before the investment results in a negative NPV. Sensitivity analysis is useful for determining consequences of specified changes in variables such as product price, sales volume, input costs and investment life span. However, only risk analysis can provide any indication of the likelihood that such events will actually occur. Management should pay special attention to controlling those variables to which NPV is particularly sensitive, once the decision has been taken to accept the proposed investment. **Break-even** analysis can be viewed as a form of sensitivity analysis, because it enables the analyst to determine the impact of changes in cost, production volume, and price on profits.

Probably, the simplest type of sensitivity analysis, barring break-even analysis, can be developed by relating the variables which influence profitability as follows:

$$P = S(p\text{-}mc\text{-}vc) - Fc$$

where:
- P = profit
- S = sales
- p = selling price
- mc = material cost
- vc = variable cost
- Fc = fixed cost

A sensitivity analysis is carried out by replacing the input variables that determine profitability with their expected values given in feasibility studies in order to determine profitability in quantitative terms.

Variable	**Best Estimate**	**Increasing Variable by 10%**	**Resulting Profit ($)**	**% Change in Profit**	**Rank Order of Variables**
Sales (unit)	10,000	11,000	120,000	20	2
Price ($/unit)	50	55	150,000	50	1
Variable costs ($/unit)					
- raw material	10	11	90,000	-10	3
- other var. costs	20	22	80,000	-20	2
Fixed costs ($)	100,000	110,000	90,000	-10	3
Profit	100,000				

The resulting profit is calculated based on 10% increase in variables, taking each variable at a time. With equal percentage change in the variables, profit is most sensitive to changes in price, followed by sales volumes and other variable costs with similar changes, and raw material and fixed costs ranking last with the same level of changes. Although this simple analysis has highlighted the most sensitive variables, no insight is provided into the probability of these changes taking place. For example, the likelihood of a 10% increase in raw material cost may be very high while the probability of a 10% increase in product price may be low.

Sensitivity analysis using discounted cash flow model is illustrated in the following example.

A sweet pomelo processing company is considering investment in a new machine to produce a new drink, and estimates of the most likely cash flow are given below.

	Year 0 ($)	**Year 1 ($)**	**Year 2 ($)**	**Year 3 ($)**
Cost of machine	-100,000			
Cash inflows:				
50,000 bottles at $2 a bottle		100,000	100,000	100,000
Variable costs		50,000	50,000	50,000
Net Cash Flows	**-100,000**	**+50,000**	**+50,000**	**+50,000**

The cost of capital is 12% and the net present value is $20,095.

Some of the variables in the above example to which sensitivity analysis can be applied are:

(i) **Sales volume.** The net cash flow will have to fall to $41,634 ($100,000/2.4019 discount factor) for the NPV to be zero, because it will be zero when the present value of the future cash flow is $100,000. The discount factors for 12% and Years 1, 2 & 3 are 0.8929, 0.7972 and 0.7118 respectively.

As the likely net cash flow is +$50,000, the net cash flow may decline by about $8,366 each year before NPV becomes zero.

Total sales revenue may therefore drop by $16,732 (assuming that net cash flow is 50% of sales). At a selling price of $2 per bottle, this represents 8,366 units. Alternatively, the sales volume may decline by 16.7% before the NPV becomes negative.

(ii) **Product price.** When the sales volume is 50,000 bottles per annum, total sales revenue can drop to $91,634 ($100,000-8,366) before the NPV becomes negative. This represents a selling price of $1.83 or a reduction of approximately $0.17 per bottle which represents about 8.4% reduction in selling price.

(iii) **Variable costs.** The total variable costs can increase by $8,366 or $0.17 per bottle. This represents an increase of 17%.

(iv) **Initial outlay.** The initial outlay can increase by the NPV before the investment breaks even. The initial outlay may therefore increase by $20,095 or 20%.

(v) **Cost of capital.** IRR for the project is nearly 30%. Consequently, the cost of capital can increase by 250% before the NPV becomes zero.

The above analysis can be summarized as follows:

Sensitivity analysis

Variable	**Switching value (%)**
Sales volume	16.7
Selling price	8.4
Variable costs	17
Initial outlay	20
Cost of capital	250

The element to which the NPV appears to be most sensitive is the selling price followed by sales volume and variable cost. It is important for the project sponsor to pay particular attention to these items so that they can be fully monitored.

Sensitivity analysis has a number of shortcomings. In particular, the method requires that changes in each key variable are isolated, but the management is also interested in the combined effect of changes of two or more variables. In addition, the method gives no indication of the likely probability of changes in key variables or of a combination of variables occurring. For instance, the sensitivity analysis may indicate that one key variable may change by 30% and another by 10%. This suggests more attention to be paid to the latter, but if the former has a probability of occurring of 0.5 and the latter 0.01, then clearly the key variable change of 30% is more important.

IX. FINANCING DECISIONS

Introduction

All investment needs to be financed, and the means of financing differ considerably between the sectors. Industrial and commercial enterprises in both the private and public sectors rely to a considerable extent on internal resources - depreciation provisions and retained earnings. Parastatal companies obtain the balance of their funds from the government as well as from financial institutions. Private industry, commercial agriculture and trade rely much more on the banks, on long-term loans and on equity capital.

The nature of funds required, however, depends on the nature of the investment, the existing capital structure of the enterprise and the cost and availability of funds of different types. Furthermore, attitudes to the appropriate financing structure vary substantially among companies, often depending on the nature of their business, the ownership of the company, the extent to which operations are conducted internationally as well as domestically, and the size of the company. This chapter looks at the financial structure of a company and the appropriate source of capital funds to finance its investment proposals.

Alternative Methods of Financing

The discussion is best introduced through a simple example. Let us assume that a company has the following assets and liabilities at its current level of operations. Fixed assets $30,000; current assets $10,000; current liabilities $5,000; equity and loan capital $35,000. The company wishes to undertake a new project which would involve an investment of $6,000 in fixed assets and $2,000 in current assets.

This proposal would call for funds totalling $8,000. Three alternative methods of financing are examined. In the first method, the purchase of assets is assumed to be financed by a short-term loan from the bank; in the second method, by a combination of long-term capital for fixed assets and a short-term loan for current assets; and in the third method, long-term capital will be raised to finance fixed assets plus $1,000 of current assets (assumed to be the hard **'core'** of total current assets) and short-term loan for the balance current assets of $1,000.

METHOD I

Assets	Without Project		With Project	
Fixed assets		30,000		36,000
Current assets	10,000		12,000	
Less current liabilities	5,000		13,000	
Net current assets		**5,000**		**(1,000)**
Total Assets		**35,000**		**35,000**
Financed by				
Equity and loan capital		**35,000**		**35,000**
		====		====

The use of short-term loans to finance both fixed and current assets has resulted in **net current assets** (or working capital) becoming negative. This is due to raising of short-term loan of $8,000 which increased current liabilities from $5,000 to $13,000.

METHOD II

Assets	Without Project		With Project	
Fixed assets		30,000		36,000
Current assets	10,000		12,000	
Less current liabilities	5,000		7,000	
Net current assets		**5,000**		**5,000**
Total Assets		**35,000**		**41,000**
Financed by				
Equity and loan capital		**35,000**		**41,000**
		====		====

The use of long-term funds for fixed assets ($6,000) and short-term loan for current assets ($2,000) causes an increase in combined capital from $35,000 to $41,000, and an increase in current liability from $5,000 to $7,000. The net current assets remain unchanged because the creation of current liability in the form of a short-term loan of $2,000 is matched by the addition of current assets of equal amount.

METHOD III

Assets	Without Project		With Project	
Fixed assets		30,000		36,000
Current assets	10,000		12,000	
Less current liabilities	5,000		6,000	
Net current assets		**5,000**		**6,000**
Total Assets		**35,000**		**42,000**
Financed by				
Equity and loan capital		**35,000**		**42,000**
		====		====

The whole of fixed assets ($6,000) plus portion of current assets ($1,000) have been met from long-term capital with the result that the net current assets improve from $5,000 to $6,000.

Financing Strategy

The use of Method I results in negative net current assets at the end of Year 1 and a substantial short-term borrowing exposure. Method II restructures the capital deficit over a longer time period and, at the end of Year 1, results in positive net current assets. Method III, recognizing the existence of **hard core** current assets, finances this element by long-term capital. **What is the appropriate method?** The answer cannot be given without considering individual circumstances. But the general principle is one of matching assets and liability maturities so that cash flows generated by an asset will be sufficient to service capital within the asset's useful life. Capital of maturity shorter than the asset's life is considered more risky since there is some possibility that the asset will not have generated adequate cash flows by the maturity date to repay debt, or reward investors. Normally one would expect long-term sources of funds to finance long-term assets and short-term sources to fund short-term assets, as our Beta Ltd. financing shows.

In practice, short-term loans can be rolled over, and the ensuing risks will have to be evaluated. Because of the generally lower average costs of short-term loans, enterprises may be tempted to finance fixed assets with short-term loans and earn more profits. Although the availability of short-term loans may continue over a long term if loans are rolled over, the fact that interest rates are liable to vary means that the borrower is vulnerable to sharp changes in borrowing costs. If short-term interest rates are expected to increase and to persist, then it may benefit a borrower to finance net current assets by long-term loans issued at current relatively lower rates. Thus, decisions about the firm's appropriate mix of short-term and long-term funds must also incorporate the consequences for profits arising from a given financing strategy.

Equity and Loan

Having established the need for long-term finance, the next step is to decide on the proportion of debt and equity capital. Lenders generally set limits on a firm's gearing. An issue that naturally arises is whether or not an ideal capital structure can be defined. The answer is unfortunately "no". The most suitable capital structure for any project must depend on the particular circumstances. At this level of generalization, the best that can be done is to indicate factors that influence the financial structure of the project. These include:

(1) **availability of finance;**

(2) **cost of financing;**

(3) **control;**

(4) **growth and stability of sales;**

(5) **asset structure (i.e. availability of assets to mortgage);**

(6) **meeting fixed commitments;**

(7) **risk.**

The riskier a project is, the better it is for it to be financed from equity. Companies that use raw materials or deal in finished products whose prices are likely to fluctuate widely (e.g. rubber, sugar, oilseeds) would require much higher equity financing Similarly, livestock and poultry producers who are likely to face serious risk of loss due to diseases, must have a sound equity base.

At a more general level, it can be stated that equity finance is suited to the early life of an investment, when returns are **uncertain**, to very **risky investments** and to investments with **highly variable income.** More specifically three special types of investment may be distinguished. First, projects involving large amounts of capital expenditure to create intangible assets. These will need equity finance. The main example of this is capital expenditure on technical assistance, advertising or other sales promotion or on extension and research. The second type of investment is that of cost-saving capital projects. These projects normally involve very little change in working capital (hence that proportion of capital which can normally be readily financed by short-term loan is absent) and they may also involve the writing off of existing assets which these projects replace. The third type of project may be typified by marketing projects. These frequently involve a substantial increase in working capital (in the shape of stocks and trade debtors) which will support an exceptionally large proportion of debt capital.

Cost of Capital

Under normal circumstances, a company would be expected to provide a lower real return to the suppliers of loan capital than to the suppliers of equity capital, the reason being that, for a given company, loan capital is the less risky investment. Payment of loan interest ranks before payment of equity dividends each year, and repayment of loan capital ranks before repayment of equity capital in the event of liquidation. Moreover, the effective cost of loan capital to a company is likely to be even further below the cost of equity capital because loan interest - unlike dividends - may attract company tax relief.

As companies increase the proportion of loan capital relative to equity capital, they gain the cost advantage of making use of relatively cheap capital but there is also a possible cost disadvantage. The payment of debt interest is a contractual obligation which must be discharged before dividends can be paid; thus the suppliers of equity capital are likely to take an adverse view of increased gearing, as it increases the risk of insufficient funds being available to pay dividends after the higher level of loan interest has been paid. In order to compensate for this additional **financial risk,** the suppliers of equity are likely to require a higher return on their capital; the cost of equity capital to a company is thus raised.

However, on balance, it is generally believed that as long as companies do not become too highly geared, the advantage is likely to outweigh the disadvantage. This net advantage will accrue to the owners of equity capital. In short, private sector companies can and do bring benefit to their owners by combining both equity and loan in their capital structures as the illustration given below shows. However this is true only in situations where the interest rate on loan capital is lower than the rate of return on investment.

An Illustration

Take a project that costs \$100,000 and which generates a net cash flow from operations (net profit before interest and depreciation) of \$10,000 in Year 1 and \$20,000 in Year 2. Other years of project life are ignored.

The project can be financed in several ways. The results of financing by three different possible combinations of equity and loan are shown below:

Option I	Year 1	Year 2
	. . . . $000's	
Equity	80	80
Loan	20	20
Total capital	100	100
Project net cash flow	10	20
Less interest 10% of $20,000	2	2
Equity cash flow	8	18
Rate of return to equity (%)	10	22.5

Option II	Year 1	Year 2
Equity	50	50
Loan	50	50
Total capital	100	100
Project net cash flow	10	20
Less interest 10% of $50,000	5	5
Equity cash flow	5	15
Rate of return to equity (%)	10	30

Option III	Year 1	Year 2
Equity	20	20
Loan	80	80
Total capital	100	100
Project net cash flow	10	20
Less interest 10% on $80,000	8	8
Equity cash flow	2	12
Rate of return to equity (%)	10	60

The above example reveals how the profits available to equity holders are subject to a greater degree of fluctuation in a **highly geared** company (Option III) than in a company with **low gearing** (Option I). An increase in net cash flow prior to interest by 100% in Option III has caused equity cash flow to rise by 500% whereas under Option I, the equity cash flow has increased by only 125%. Another point that emerges from this example is that high gearing is risky when profits are low as seen in Year 1 of Option III.

Finally, it shows that with enough gearing equity holders can make money **(trading on equity)** not only on the capital put in by them but also on the funds lent by the creditors. In Year 2 of Option III, for example, the equity holders are getting a return of 60% on the small amount of capital they have put up, but the loan is still earning only the interest rate that has been fixed at 10%.

Capital Structure in Public Enterprises

For the public enterprises which normally obtain all their long-term external finance from the government, it would seem pointless to have mixed capital structures similar to those found in the private sector. The government would be both the provider of equity capital and lender of debt capital, and it would also be the provider of tax relief on loan interest. Thus all the costs and benefits of gearing would accrue to the government, which would receive both the dividend and the interest, and there would be no net benefit, as the government would also be the receiver of the reduced company tax payment resulting from the tax relief on loan interest. Thus the government would bear all the risks of the public enterprise investment, however they were to be financed. Although this is so, financing by a loan may compel the enterprise to tighten management efficiency and seek economies to pay interest on loan. This is likely to instill some financial discipline.

In order to ensure that investments made in this sector produce a satisfactory economic rate of return, it would be necessary for the government to set a minimum acceptable rate of return for the total invested capital in each enterprise. This rate of return must be calculated irrespective of the proportion of the investments that is financed by borrowing, new equity, or retained earnings.

Deadweight Debt

Private firms have an interest in limiting the fixed commitment element of their capital structures, because the situation could arise where financial resources are insufficient to meet interest or capital repayment obligations, resulting in the company being forced into liquidation or capital reconstruction or at best requiring the raising of additional finance. In the latter situation, the extra finance - which is likely to be a loan with further fixed interest commitments - may be a **deadweight burden** to the company, in the sense that it is not applied to productive investment (but to meet the interest payments), and thus will not be able to generate future cash flow, from which its own interest charge can be met. This is exactly a situation to which most public enterprises are highly susceptible. These enterprises, when fully financed by debt capital from government, may face a situation (possibly caused by downturn in the industry's fortunes) where they would not be able to meet fully interest payments from internally generated cash flow, forcing them to resort to deadweight financing.

Taxation and Gearing

The incidence of taxation has its influence on the capital gearing of the companies since it encourages them to raise loan capital in larger proportions than would be normally justified.

This is because loan interest is generally an allowable deduction for tax purposes. For example, the effective cost of a 10% loan to a company making profits would be 5% assuming company tax rate of 50%.

Inflation and Gearing

In inflationary times, it is beneficial to borrow at a fixed rate of interest, since the company will only be required to pay the fixed charge. When repayments are made, the money has fallen in value. This is likely to lead companies into difficulties by over-gearing in the belief that borrowing is in itself advantageous during a time of inflation, independently of the use to which borrowed money is put. Such a belief is particularly dangerous when interest rates are high.

As the effects of inflation become more widely perceived, the lenders possibly demand a higher rate of interest as inflation compensation, and the borrower accepts that he must pay this in order to obtain his monetary gain on borrowing. If inflation is running at 12%, an interest rate of 15% consists of **2.67% pure interest** and **12.33% inflation compensation.** But the borrower receives tax relief on the total interest charge of 15% although 12.33% of it is in reality a payment for the capital gain he is building up on the loan. In this way, the borrower is being subsidized out of national taxation.

X. WORKING CAPITAL

Introduction

The concept of working capital is related to the portion of capital that is needed to operate a company's fixed assets, i.e. to finance its operating cycle (procurement→ production→ distribution). Traditionally, working capital (or net current assets) is defined as the difference between current assets and current liabilities. The concept becomes more meaningful when it is defined as the difference between long-term capital and fixed assets, which from a computational point of view is identical to the previous definition. First, it can now be said that working capital represents the financial outcome of the company's long-term policy. This is true because the amount of long-term capital is the result of the long-term financing decisions of the company and fixed assets are the result of its long-term investment decisions. Secondly, working capital represents either the portion of the company's capital which finances its current assets when working capital is positive, or the portion of its current liabilities which finances its fixed assets when working capital is negative. In other words, working capital is associated with the strategy of the company with respect both to its long-term investment decisions and to some aspects of its financial risk management.

Computing Working Capital Requirements

General method of computation. There can be no set formula for calculating working capital requirements for all enterprises. Each project has to be looked at by itself taking into account the relationship between the industry on the one hand and its suppliers and customers on the other.

The normal practice is to calculate some of the main requirements of working capital in terms of annual production or sales.

- If suppliers of raw materials give two months' credit, then the project can get credit for 2/12 of its annual purchases.

- If a firm extends one month's credit to its customers, then project must be able to finance 1/12 of the annual cost of sales.

- If materials need to be held in stock for an average of 3 months the project must be able to finance three months' purchases. For stocks of finished goods, a similar approach is used.

- Funds for spares are based on a certain percentage of the cost of plant and equipment.

- Requirements for additional cash as working balances to be kept in hand are based on cash budget or calculated as a percentage of total operating expenses.

A typical format for a working capital projection is given below.

	Year 1	Year 2	Year 3	Year 4	Year 5	Year 6
	.... ($'000)					
Current Assets						
Stocks	100	110	120	130	130	-
Trade debtors	40	44	50	60	60	-
Cash	20	22	25	25	25	-
Total current assets	**160**	**176**	**195**	**215**	**215**	-
Less **Current Liability:**						
Trade creditors	80	100	100	110	110	-
Bank overdraft	25	10	5	-	-	-
Total current liability	**105**	**110**	**105**	**110**	**110**	-
Working capital	**55**	**66**	**90**	**105**	**105**	-
Incremental working capital	**55**	**11**	**24**	**15**	-	(105) 1/

1/ At the end of the project, subject to losses in the value of stocks or of debtors, incremental working capital is fully recovered.

Other methods of estimation. There are other methods, of which **percent of sales method** is important for enterprises which have been operating for some time and display a direct relationship between sales and working capital. This relationship should be stable for projections of working capital based on sales (a certain percentage) to be realistic.

Working Capital Forecast in a Manufacturing Company

A general issue is raised as to what will be the working capital requirements to finance a new project? To show the usual requirements in simple form, the following items are tabulated conforming to the operating cycle.

(1) The cost of raw materials, wages and overheads.

(2) The period during which raw materials will remain in stock before issue to production.

(3) The period during which the product will be processed through the factory.

(4) The period during which finished goods will remain in the warehouse.

(5) The lag in payment to suppliers of raw materials and services.

(6) The lag in payment to employees.

(7) The lag in payment by debtors.

Most of these points will be included in a computation of working capital requirements. An illustration is given below.

Data. A company plans to produce 60,000 units of furniture. The selling price of a unit is $5. The expected ratios of cost to selling price are raw materials 60%, direct wages 10%, production overheads 20%. Raw materials (timber) are expected to remain in store for an average of 2 months before issue to production. Each unit of production is expected to be in process for 1 month. Finished goods will stay in the warehouse awaiting despatch to customers for about 3 months. Credit allowed by creditors is 2 months from date of delivery of raw materials. Credit given to debtors is 3 months from the despatch of goods. There is a regular production and sales cycle.

Forecast of Working Capital Requirements

	Period	Total	Raw materials	Work in progress	Finished Goods	Debtors	Creditors
	(months)	. . . .	. . . .	.($) . . .	. . . .	. . . . .	. . . .
1 Materials							
(a) in stock	2		30,000				
(b) in work progress	1			15,000			
(c) in finished goods	3				45,000		
(d) credit to debtors	3					45,000	
	9						
(e) credit from creditors	2						30,000
Total	7	105,000					
2 Wages							
(a) in work in progress	1			2,500			
(b) in finished goods	3				7,500		
(c) credit to debtors	3					7,500	
Total	7	17,500					
3 Overheads							
(a) in work in progress	1			5,000			
(b) in finished goods	3				15,000		
(c) Credit to debtors	3					15,000	
Total	7	35,000					
4 Profit							
(a) credit to debtors	3					7,500	
Total	3	7,500					
Grand Total		165,000	30,000	22,500	67,500	75,000	30,000

It will be observed that this forecast includes elements of cost and profit, and excludes the cash position. Cash is an important element of working capital, so the cash position ascertained from the cash budget (say $10,000) must be added to give the final figure for working capital required.

	. . $ . . .
Requirements as shown in forecast	165,000
Cash as per cash budget	10,000
Total working capital required	**175,000**

Working Capital Forecast in a Trading Enterprise

This forecast is usually prepared with sales as the key independent variable. An illustration is given below showing the methodology as in the previous example.

The following is an extract from the draft balance sheet of Ceylon Coffee Ltd. on 31 December 1989.

	. . . . . ($)	. . . .
Fixed Assets		**50,000**
Current Assets		
Stock	10,000	
Trade debtors	5,000	
Sub-total	**15,000**	
Less **Current Liabilities**		
Trade creditors	6,000	
Bank overdraft	4,000	
Sub-Total	**10,000**	
Working capital		5,000
Total Assets		**55,000**

It is estimated that the sales for 1990 will increase by 50%. The bank has refused to allow the overdraft limit of $7,000 to be exceeded and the company has decided not to incur capital expenditure in 1990. Net profits are estimated at $10,000 for 1990, all of it to be invested in a subsidiary company. Both credit periods and stock turnover are expected to remain constant.

You are required to forecast working capital for 1993.

Working Capital Forecast as at
31 December 1993

	..... $	
Current Assets		
Stock	15,000 1/	
Trade debtors	7,500 1/	**22,500**
Current Liabilities		
Trade creditors	9,000 1/	
Bank overdraft	7,000 2/	**16,000**
Estimated working capital		**6,500**
Amount required:		
Forecast working capital		6,500
Present working capital		5,000
Additional		**1,500**

1/ As sales are to rise by 50%, related variables of stock, debtors and creditors are increased correspondingly.
2/ Overdraft is drawn to the limit.

Working Capital and IRR Calculation

Working capital can be an important variable in IRR calculations. An example is given illustrating this point.

An Illustration. A company is assessing a project to extract oil from oilseeds. On the basis of the following information, make a recommendation as to the feasibility of the proposal in financial terms.

(1) Plant and machinery will cost $200,000 and will last 5 years at the end of which time it will have a scrap value of $40,000.

(2) The factory premises will be rented at a cost of $1,000 per annum.

(3) Sales will be:

Year 1	500 tons
Year 2	1,000 tons
Year 3	1,000 tons
Year 4	1,500 tons
Year 5	1,000 tons

(4) The selling price per ton will be $180.

(5) Variable production costs will be $60 per ton of output made up as follows:

Materials - $30)
Labour - $10) per ton
Overheads - $20)

(6) Fixed costs will amount to $39,000 per annum including depreciation on a straight line basis.

(7) Working capital requirements will be as follows:

(a) two months' supplies of raw material will be held;
(b) debtors will be allowed on average 2 months' credit while 1 month's credit will be received from the suppliers of raw materials;
(c) stock of finished goods (in tons) will be held equivalent to 10% of the year's sales.

(8) The company tax rate is 40% of net profit and is paid 1 year in arrears.

(9) The company's borrowing rate is 15%.

Financial Rate of Return Involving Working Capital
Suggested Solution

The project is acceptable as its net present value (at 15% discount rate) is $71,953. The derivation is shown below.

	Year 0	Year 1	Year 2	Year 3	Year 4	Year 5	Year 6
Cash Inflows							
Profit before depreciation (see Schedule 2)	-	52,000	112,000	112,000	172,000	112,000	-
Cash Outflows							
Investment in plant & machinery	200,000	-	-	-	-	(40,000)	-
Investment in working capital (see Schedule 1)	19,250	19,250	-	19,250	(19,250)	(38,500)	
Tax	-	-	8,000	32,000	32,000	56,000	32,000
Total outflows	219,250	19,250	8,000	51,250	12,750	(22,500)	32,000
Net cash flow	(219,250)	32,750	104,000	60,750	159,250	134,500	(32,000)
PV factor at 15%	1	0.37	0.756	0.658	0.572	0.497	0.432
Present Value	(219,250)	28,492	78,624	39,974	91,091	66,846	(13,824)

Net Present Value = $71,953

FRR = 26%

SCHEDULE 1

Incremental Working Capital

As stocks and debtors will arise at the beginning of the first year, the initial investment in working capital is treated as arising at time '0'. Working capital changes arising from changes in sales volume in Years 2, 4, and 5 are therefore treated as occurring at the end of the previous year. The investment in working capital is recovered or released at the end of Year 5 on termination of the project. The working capital requirements for the various levels of sales are:

Sales (tons)	500	1,000	1,500
	....	$	
Finished stock at $60/ton 1/	3,000	6,000	9,000
Raw material stock (2/12 of the year's purchase of materials)	2,500	5,000	7,500
Debtors (2/12 of sales)	15,000	30,000	45,000
	20,500	**41,000**	**61,500**
Less: Creditors 1/12 of material cost	1,250	2,500	3,750
Working capital	**19,250**	**38,500**	**57,750**

Annual sales (tons)	500	1,000	1,000	1,500	1,000
Working capital ($)	19,250	38,500	38,500	57,750	38,500
Incremental	19,250	19,250	-	19,250	(19,250)

1/ Fixed costs are not included and 1/10 of year's sales at $60/ton would be (50 x 60), (100 x 60) and (150 x 60).

The initial investment at time '0' is \$19,250 and further investments of \$19,250 are required at the end of Years 1 and 3. There will be a reduction of \$19,250 in Year 4 and the remaining \$38,500 will be released at the end of Year 5.

SCHEDULE 2

Incremental Profits at Different Levels of Sales

Sales (tons)	500	1,000	1,500
	....	\$	....
Sales revenue	90,000	180,000	270,000
Variable Production Costs:	30,000	60,000	90,000
Fixed overheads			
Lease	1,000	1,000	1,000
Others (of which depreciation \$32,000)	39,000	39,000	39,000
Total costs	**70,000**	**100,000**	**130,000**
Incremental profit	20,000	80,000	140,000
Add back depreciation 1/	32,000	32,000	32,000
Incremental profit in cash	52,000	112,000	172,000

1/ $\frac{200,000 - 40,000}{5}$ = \$32,000

Inflation and Working Capital

Prices of many agricultural products are controlled by the government and when a price increase is allowed, it does not always match the increase in cost. For example, let us take a feedmill company which has faced an increase in the cost of its products by 30%, and other costs by 10%. The government has allowed it to increase selling price by 20%. Profit and loss statement, working capital statement and a cash flow are given below to illustrate the effects of cost and price increases.

Inflation Effect on Profit and Loss

	Current Position 1/	Annual Inflation Rate		With Inflation
	..................$	....		
Sales	10,000	20%		12,000
Stocks purchase	5,000	30%		6,500
Gross margin	**5,000**			**5,500**
Depreciation	1,000	-	1,000	
Other costs	2,000	10%	2,200	3,200
Net Profit	**2,000**			**2,300**
Net operating cash flow (Profit and depreciation)	3,000			3,300

1/ Current position data are assumed figures; others are derived.

Inflation Effect on Working Capital

	Current Position 1/	Increase due to Inflation	With Inflation
	..................$	....	
Trade debtors	6,000	20%	7,200
Stock	4,000	30%	5,200
	10,000		**12,400**
Less trade creditors	5,000	20%	6,000
Working capital	**5,000**		**6,400**

1/ Current position data are assumed figures; others are derived.

Inflation Effect
on
Source and Application of Funds

	$
Incremental Sources	
Net profit	300
Trade creditors	1,000
	1,300
Incremental Application	
Stocks	1,200
Trade debtors	1,200
	2,400
Net cash outflow	**1,100**

As shown above, the working capital after inflation has increased and the company has a net cash outflow of $1,100, funds for which need to be raised externally if the value of the company is not to decline in real terms.

XI. FINANCIAL STATEMENT STRUCTURE ANALYSIS

The essential features of a set of accounts have been introduced in Chapter IV; the purpose of this Chapter is to build on this foundation. Balance sheets are analysed first, followed by profit and loss statements, source and application of funds statements and value added statements; problem areas in the measurement of assets, liabilities and income are dealt with later, in Chapter XIV.

Balance Sheet

Structure

The structure of the balance sheet (or for that matter of the profit and loss statement or funds statement) is determined by the need to reflect **a true and fair** view of the state of affairs of an enterprise. To achieve fairness in reporting, the balance sheet is prepared in accordance with established practices, accounting standards and legal requirements. A real understanding of the statement requires a thorough knowledge of the composition and contents of the accounts and their valuation.

The balance sheet terms used such as share capital, reserves, current liabilities, long-term liabilities, fixed assets and current assets are normally appropriate headings for summaries and classifications. Other headings should be used if it is considered that they will better reflect the company's financial state. Let us look at a typical balance sheet of an enterprise.

Sugar Products Ltd.
Balance Sheet as 31.12.90

Liabilities & Equity		$	Assets			$
			Fixed Assets			
Share capital		100,000	Land & building			
(10,000 shares of $10 each)			at valuation	130,000		
			Less depreciation	30,000	100,000	
Share premium reserve		54,000	Plant machinery			
Revaluation reserve		6,000	& equipment at cost	90,000		
			Less depreciation	20,000	70,000	
Revenue reserves		7,000	Motor vehicles	15,000		
			(at cost)			
			Less depreciation	5,000	10,000	180,000
			Goodwill			10,000
Share capital		167,000	Investments			
& reserves			Investment in		64,000	
			subsidiaries at cost			
Long-term Liabilities						
10% Debentures	60,000		Investment in		11,000	75,000
Bank loan	55,000	115,000	associated companies at cost			
			Current Assets			
			Raw materials		10,000	
			Work in progress		6,000	
			Finished goods		15,000	
			Accounts receivable		15,000	
Current Liabilities:			Cash on hand & at bank			
Accounts payable	8,000		prepaid expenses		5,000	51,000
Proposed dividend	10,000					
Provision for company	12,000		Expenditure Carried Forward			
tax			Preproduction			
Bank overdraft	10,000	40,000	expenditure			6,000
		322,000				322,000
		=======				=======

The information presented above relates to the following.

Fixed assets, which are items of value: (a) held by an enterprise for use in production or supply of goods and services, for rental to others, or for administrative purposes; (b) acquired or constructed with the intention of being used on a continuing basis; and (c) not intended for sale in the ordinary course of business. Fixed assets, which would include long-term investment, can be either tangible or intangible.

Tangible fixed assets. Tangible fixed assets will normally include land and buildings, leasehold property, plant, machinery and equipment, vehicles, tools, long-term assets obtained under a hire-purchase or instalment payment agreement, and equipment rented out on a regular basis to earn revenue. These assets are generally carried at cost, less provision for depreciation which is shown separately on a cumulative basis. Most fixed assets should be depreciated on a rational and systematic basis over the life of the asset. In response to specific changes in prices, it is permissible to revalue property, plant and equipment. If revalued, unrealized surplus arising from an appraisal or re-appraisal should be dealt with (or credited) under the heading of revaluation reserve.

For example, land and building has been revalued in Sugar Products Ltd. and the surplus of $6,000 is shown under revaluation reserve. Deficits on revaluation should be charged (or debited) to revaluation surplus unless there is insufficient surplus arising from previous revaluations of the same class of assets, in which case the balance should be charged to the profit and loss account for the year.

Intangible fixed assets. These are assets without physical characteristics, and whose value lies in the rights, privileges and competitive advantages which they give the owner. Examples of intangible assets are goodwill, patents, copyrights, licences, franchises and trademarks. If material, the major items among the intangible assets should be shown separately, as in the case of "goodwill" in the Sugar Products Ltd. Intangible assets acquired for cash or its equivalent are recorded at the value of amounts paid. Where an intangible asset has been acquired otherwise, for example by issue of shares, the basis of valuation should be fully described as for example, 'at cost being the values assigned to the shares issued therefor'. Intangible assets having a limited useful life should be amortized over that life and this fact should be disclosed.

Long-term investments. Investments are assets not directly identified with the operating activities of an enterprise. They occupy a subsidiary role to the principal revenue producing activities of the company and are expected to contribute to the success of the enterprise either by exercising certain favourable effects upon sales and operations generally, or by making an independent contribution to earnings over the long-term. Investments other than temporary investments should be segregated so as to distinguish each of the following major groups: investment in associated companies; investment in subsidiaries; investment in related companies; and other investments. Any investment acquired should be recorded at cost at date of acquisition. The latter is measured by the purchase price of the security or the fair value of the asset given up in exchange. The basis of valuation of each group must be indicated. The market value of quoted securities should be disclosed in all cases. Where the book value of any investment is significantly different from the value of the underlying assets, additional information is desirable especially where the investment forms a significant proportion of the investing company's total assets.

Current assets. These are items of value ordinarily realizable within one year from the date of the balance sheet or within the normal operating cycle where this is longer than a year. The operating cycle of an enterprise refers to the average time lag between purchase of materials entering into the production process and the final cash realization.

Such assets include inventories (raw materials, work in progress and finished stock), trade and other debtors, bills receivable and pre-paid expenses, quoted and other readily realizable investments (other than trade investments, investment in subsidiaries and associated companies, and other investments intended to be held continuously even though they may happen to be quoted or are otherwise readily realizable); cash and bank balances.

Inventories are valued at the lower of historical cost and net realizable value and the accounting policies adopted for the purpose of valuation, including the cost formula used (e.g. First in First out (FIFO), Last in First out (LIFO), or weighted average cost), should be disclosed. Trade and other debtors should be segregated. If material, trade debtors should be sub-divided as between trade debtors, bills receivable, hire-purchase debtors, etc. Trade debtors would normally include all debts arising from sales or revenue earning activities. Loans and advances to directors should be separately disclosed. The provision for the allowance for doubtful accounts, if any, should be separately disclosed and deducted from the total of the balance to which it applies. Investments classified as current assets are valued at the lower of the cost and net realizable value. Cash and bank balances include cash on hand and current accounts, and short-term deposits with banks. If there are material amounts of assets in overseas countries, their value should be converted at the exchange rate ruling on the balance sheet date (closing rate).

Deferred charges as distinct from prepaid expenses should be shown separately under a description such as expenditure carried forward. Pre-production expenditure of $6,000 in Sugar Products Ltd. is such an example. The amount written off in the year should be separately disclosed in the profit and loss account.

Current liabilities. These include all liabilities payable at the demand of the creditor or within one year from the date of the balance sheet, or within the normal operating cycle where this is longer than a year. Such liabilities include trade creditors (including bills payable), other creditors, short-term loans, bank overdrafts, taxes, proposed dividends, provisions for specific revenue commitments and contingencies and the current portion of long-term liabilities. Trade creditors would normally include all items relating to the trading activities being of a revenue nature. Amounts owing on loans from directors, amounts owing to a holding company or subsidiaries and amounts owing to associated companies should be shown separately. Contingent assets and contingent liabilities should be disclosed and quantified if possible.

Long-term liabilities. These are amounts owed by an enterprise to outside parties and due for settlement after 12 months from the date of the balance sheet or after the normal operating cycle where this is longer than a year. Long-term liabilities include bonds, debentures, mortgage loans, long-term loans from financial institutions, from affiliated companies, officers or shareholders. The following items should be disclosed separately - secured loans, unsecured loans, intercompany loans and loans from associated companies. A summary of the interest rates, repayment terms, covenants, subordinations and conversion features should be shown. Accrued interest in respect of a secured liability, which is itself secured, should be included with the liability. Otherwise interest should be included in current liabilities.

Share capital and reserves, which represent the amounts owed to the shareholders by an enterprise and usually not due for settlement until the termination of the enterprise.

Share capital description must show authorized amount, par value, number of shares issued and paid up according to different class of shares (e.g. ordinary shares and preference shares). **Reserves** should be distinguished between: capital reserves, i.e. those which for statutory reasons or because of the provisions of the Memorandum of Association or Articles of Association of a company or for other legal reasons, are not free for distribution through the profit and loss account, for example, revaluation reserve and revenue reserves which are fully available for distribution. Where a company issues shares for which a share premium is received by the company, the amount of that premium should be shown in the balance sheet. Additions to or withdrawal from **revenue reserves** (including retained profits) should be reflected in the profit and loss account. All transfers between reserves, and transfers from reserves to share capital (e.g. bonus issue of shares) should be disclosed.

Adverse balance in profit and loss statement. Available reserves should be appropriated to extinguish an adverse balance on profit and loss account. Where this practice is not adopted, the adverse balance should be grouped with and be a deduction in arriving at the sub-total of share capital and reserves. Preferably such deduction should be made first from revenue reserves; if these are insufficient, then from the aggregate of revenue and capital reserves. If both are insufficient, the adverse balance should be deducted from the aggregate of share capital and reserves. If the total of capital and reserves fails to cover the amount of accumulated losses, the balance should be shown as a bracketed figure describing as 'excess of accumulated losses over the total of capital and reserves'.

Presentation

The assets are listed in increasing order of liquidity, while the liabilities and capital are ordered according to the remoteness with which they are likely to be paid.

To permit adequate identification of the separate assets and liabilities of the enterprise, the amount at which a particular asset or liability is stated, is generally not reduced by deducting another liability or asset. However, such off-setting might be appropriate when a legal right of set-off exists and the off-setting represents the expectation as to the realization or settlement of the asset or liability. Progress payments and advances may be deducted from the amount of related construction work in progress, provided disclosure is made in the accounts.

Profit and Loss Statement

Structure

The profit and loss statement may be sub-divided into:

- manufacturing account showing prime cost of production;
- trading account showing cost of sales;
- profit and loss account showing net profit; and
- profit and loss appropriation account showing the distribution of profits.

A simplified example is given below to illustrate the above structure.

Profit and Loss Statement for Year ended 31.12.90

			($'000)
Sales			6,138
Cost of sales			4,870

Materials consumed	1,000	
Wages	1,200	
Other direct expenses	300	
Prime cost		2,500
Production overhead		
Indirect labour		100
Power, light and heat		50
Rent and rates		20
Depreciation		707
Other expenses		10
Cost of goods manufactured		3,387
Opening stock of finished goods		500
Purchases of finished goods		1,800
Less: closing stock of finished goods		817
		4,870

Gross profit (margin)		1,268
Marketing costs	260	
Administration expenses	90	
Research and development expenses	51	401
Profit on ordinary activities before interest and tax		867
Export contribution (other income)		885
Profit before interest and tax		1,752
Interest payable		382
Profit before tax		1,370
Tax		470
Profit after tax		900
Extraordinary items 1/		23
Profit for the financial year		923
Balance brought forward from previous year		100
Dividends paid and proposed		300
Profit retained for the year		723

1/ Extraordinary income less extraordinary charges after tax.

Manufacturing account. The first part of the manufacturing account, which establishes the **prime cost** of production comprises the following three items:

- cost of direct materials consumed;
- direct wages; and
- direct expenses.

The prime cost includes only those materials which actually become part of the finished product and only wages of those who work directly on the product (as opposed to those who work in supervising and administering such workers).

The second part of the manufacturing account consists of **production overheads.** These are also known as 'manufacturing overheads', 'factory overheads' or 'works overheads' and the main constituent elements are rent, rates, insurance, lighting, heating and power of the factory; depreciation, repairs and maintenance of plant and machinery and indirect wages of such staff as supervisors, storemen and inspectors who work in the factory but not directly on the product. With some expenses such as rent and rates, it may be necessary to apportion part to administration overheads to be charged to the profit and loss account proper, and part to factory overheads to be charged to manufacturing account.

The third part of the account shows the adjustment in respect of partly finished goods or 'work in progress'. The cost of any work in progress brought forward from the previous period is added to the total manufacturing costs for the period and the cost of any work not completed at the end of the period is deducted from the total. The resultant figure is the production cost of the finished goods available for sale which is then transferred to the trading account.

Trading account. This account shows the cost of goods transferred from manufacturing account, and of direct purchases of finished goods from outside along with associated expenses such as handling and transportation, which, together with the cost of any finished goods brought forward from the previous period, make up goods available for sale. From this total is deducted the cost of any goods unsold to give the cost of sales, which is matched with the sales for the period. The resulting figure is gross profit (or gross loss). The gross profit/loss is transferred to the profit and loss account.

Profit and loss account. Marketing and distribution costs, administrative costs and any research and development expenses are deducted from the gross profit (or gross loss) to determine the company's **operating profit** for the period. To this, investment income is added and interest payable deducted to give profit before taxation, from which company tax is deducted to give the net profit or loss for the period. The net profit or loss will finally be adjusted for extraordinary income or charges to show all inclusive profit/loss.

The profit and loss appropriation account. This account indicates the profit available for distribution to the shareholders in the form of dividends, and for transfer to reserves to be carried forward to the next period. The profit available for appropriation is represented by the balance brought forward from the previous period plus or minus the net profit or loss for the period. **Retained profit (or earnings)** represent a residual element. The existence of retained profit does not imply funds available to shareholders. Such availability depends on the firm's liquidity.

Main Uses

The profit and loss account provides a measure of return from the business as well as its ability to meet financial obligations such as debt payments, rent, wages and other expenses during the year. The statement thus reveals the success or failure of a business over time as well as the costs and returns associated with the use of varying amounts of capital and credit. The information in the statement, combined with that in the balance sheet, allows the shareholders to judge the performance of management.

Statement of Source and Application of Funds

Structure

As indicated in Chapter IV, the statement of source and application of funds or simply funds statement is prepared from financial data generally identifiable in the income statement, balance sheet and related notes. The "funds statement" can be prepared in a variety of forms, depending on whether it is desired to emphasize changes in cash, net liquid funds, total financing, working capital movements or some other particular aspect. Let us consider the example given in the form of proforma statement.

PROFORMA

Statement of Source and Application of Funds
(for the Year ended 31.12.90)

		($'000)
Source of Funds		
Profit before tax		100
Adjustments for items not involving movement of funds		
Depreciation and amortization		59
Total generated from operations		**159**
Funds from other sources:		
- Long-term loans		6,000
Total inflow of funds		**6,159**
		===
Application of Funds		
Dividends paid	-	
Tax paid	-	
Purchase of fixed assets	5,841	
Repayment of loans	-	
Payment for technical assistance	200	6,041
Net increase in funds		**118**
		===
Changes in working capital		
Increase in inventories	510	
Increase in receivables	468	
Increase in payables	(95)	
Decrease in cash and bank balances	(765)	**118**
		===

Funds provided from or used in the operations of an enterprise are normally shown separately in the statement. Other sources of funds are also stated separately, these include for example:

(a) proceeds from sale of long-term assets;

(b) issue of long-term debt ($6 million in the above example);

(c) issue of shares for cash or other assets.

A distinction is also made in the statement between applications arising from changes in current assets and current liabilities and purchases of fixed assets, repayment of loans, payment of tax and dividends.

Different forms of presentation can be used to show the amount of funds provided from or used in the operations of an enterprise. A method commonly used is to show the net profit (or loss) and to make adjustments for those revenues or expenses that do not involve a movement of funds in the current period, for example, depreciation. An alternative method is to begin with revenues that provided funds during the period and deduct the costs and expenses that involve a movement of funds. The resulting amount is defined as funds from operations.

Uses of Statement of Source and Application of Funds

The analysis of this statement gives us a real insight into the financial operations of a firm - an insight that is valuable in analysing past and future expansion plans of the firm and their impact on liquidity.

The analyst can detect imbalances in the uses of funds and recommend appropriate actions. For example, an analysis spanning the past several years might reveal a growth in inventories out of proportion with the growth of other assets and with sales. Upon analysis, he might find that the problem was due to inefficiencies in inventory management.

Another use of the statement is in the evaluation of the firm's financing. An analysis of the major sources of funds in the past reveals what portion of the firm's growth was financed internally and what portion externally. In evaluating the firm's financing, the analyst will wish to evaluate the ratio of dividends to earnings relative to the firm's total need for funds. Funds statements are useful also in judging whether the firm has expanded at too fast a rate and whether financing is strained. For example we can find out whether trade credit (suppliers' credit) has increased out of proportion to increases in current assets and to sales. It is also revealing to analyse the mix of short and long-term financing in relation to the funds needs of the firm. If these needs are primarily for fixed assets and permanent increases in working capital, we might be concerned if a significant portion of total financing came from short-term sources.

An analysis of a projected statement of sources and application of funds would assist the firm in planning medium and long-term financing. The statements would disclose the firm's prospective need for funds, the expected timing of these needs and their nature - that is whether the increased investment is primarily for inventories or fixed assets. Given this information, along with the expected changes in accounts payable and the various accruals, the financial manager can arrange the firm's financing more effectively.

A Case for the Cash Flow Statement

Users of financial statements are interested in the liquidity, viability and financial flexibility of an enterprise. A working capital-based funds flow statement presents information which helps in the assessment of these factors but, by concentrating on the movement in working capital rather than cash flow, such a statement omits important aspects of cash flow and may obscure movements in those critical components within the funds such as cash balances, debtors and creditors. The other advantages of the cash flow statement are:

(i) Cash flows can be a direct input into a business valuation exercise and, therefore, historical cash flows may be directly relevant in a way not possible for funds flow data.

(ii) Funds flow data incorporating movements in working capital can obscure movements relevant to the viability and liquidity of an enterprise. For example, a potentially serious decrease in cash available may be masked by an increase in stock or debtors. Enterprises can therefore run out of cash while reporting increases in working capital available.

(iii) A funds flow statement is based largely on the difference between two balance sheets. It reorganizes data but does not provide any new data. A cash flow statement includes new data.

(iv) As cash flow monitoring is a normal feature of business life and not a specialized accounting concept, cash flow is a concept which is easier to understand than changes in working capital.

(v) Investors, formally or informally, develop a model to assess and compare the present value of the future cash flows of enterprises. Historical cash flow information could, therefore, be useful to check the accuracy of past assessments and indicate the relationship between the enterprise's activities and its receipts and payments.

Classification of Cash Flows

The cash flow statement classifies the elements of cash flow between the economic activities, operating, investing, and financing, which together make up the net movement of cash for the period. This net movement is added to the balance of cash brought forward at the beginning of the period to give the balance of cash at the end of the year. For the purposes of this statement, cash equivalents are to be treated as cash and reported as appropriate. Cash equivalents are defined as short-term, highly liquid investments which are both readily convertible into known amounts of cash and sufficiently near maturity that there is no significant risk that they will change in value in response to interest rate variations. Two examples of cash flow statements are given at the end of this chapter. The first is for an enterprise and the second for a bank.

Operating cash flows including returns from investing and financing. Flows from operating activities are generally the effects of transactions and other events relating to operating or trading activities together with returns from investing and financing. Operating cash flows include:

(a) cash receipts from sales of goods or services, including receipts from the collection or sale of debtors, gross value added tax or other sales tax levied (including both short and long-term receipts from customers arising from those sales);

(b) cash receipts from interest, dividends and other income arising from investing and financing activities. The cash flows of the principal amounts are dealt with under investing and financing as appropriate;

(c) cash refunded from customs, or any other tax authority, in relation to value added tax (VAT), or other sales tax on operating activities;

(d) cash receipts from the relevant tax authority of tax rebates, claims or overpayments which relate to transactions whose cash flows are classed as operating;

(e) all other cash flow receipts not covered by (a)-(d) above which do not stem from transactions defined as investing or financing activities. Any miscellaneous cash receipt which is included under this heading should be shown separately if it is material or if it provides information of a significant value about the activities of the enterprise.

Cash outflows for operating activities are:

(a) cash payments to suppliers for goods or services, including any VAT or other sales tax levied, and short and long-term payments to suppliers arising from those purchases;

(b) cash payments to and on behalf of employees;

(c) cash payments of interest to lenders and other creditors;

(d) cash paid to customs, or any other tax authority, in relation to value added tax or other sales tax on operating activities;

(e) cash paid to the relevant tax authority of tax due on transactions whose cash flows are classed as operating;

(f) all other cash payments not covered by (a)-(e) above which do not stem from transactions defined as investing or financing activities. This might include fines or penalties and charitable donations. Any miscellaneous cash payment which is included under this heading should be shown separately if it is material or if it provides significant information about the activities of the enterprise.

Investing Cash Flows

The cash flows included in investing activities are generally those related to the acquisition and disposal of any asset which is held as a fixed asset.

Cash inflows from investing activities are:

(a) receipts from collections or sales of loans made by the enterprise and of other entities' debt (other than cash equivalents) which were purchased by the enterprise;

(b) receipts from sales of investments in other enterprises;

(c) receipts from sales of land and buildings, plant and machinery and other fixed assets;

(d) cash refunded by customs, or any other tax authority, in relation to value added tax or other sales tax, on investing activities;

(e) cash receipts from the relevant tax authority of tax rebates, claims or overpayments which relate to transactions whose cash flows are classed as investing.

Cash outflows for investing activities are:

(a) disbursements for loans made by the enterprise and payments to acquire debt instruments of other entities (other than cash equivalents).

Financing Cash Flows

The cash flows included in financing activities are generally those relating to the manner in which the other activities of the enterprise are financed.

Cash inflows from financing activities are:

(a) proceeds from issuing shares or other equity instruments;

(b) proceeds from issuing debentures, loans, notes, bonds, mortgages and from other short or long-term borrowing;

(c) cash receipts from the relevant tax authority of tax rebates, claims or overpayments, which relate to transactions whose cash flows are classed as financing;

Cash outflows in respect of financing activities are:

(a) payment of dividends;

(b) other distribution to owners, including outlays to reacquire or redeem the enterprise's share;

(c) repayments of amounts borrowed;

(d) cash paid to the relevant tax authority of tax due on transactions whose cash flows are classed as financing

Example 1[1/]

This example gives the illustrative cash flow statement of a company XYZ Ltd.

Cash flow statement for XYZ Ltd. for the year ended 31 December 1990

	.... (£'000) ...	
Operating Activities:		
Cash received from customers	285	
Interest and dividends received	25	
Cash payments to suppliers	(190)	
Cash paid to and on behalf of employees	(85)	
Interest paid	(10)	
Payments to Customs & Excise on operating activities	(15)	
Corporation tax paid	(8)	
Charitable donations	(5)	
Cash flow from operating activities		**(3)**
Investing Activities:		
Purchase of shares in ABC Ltd. (trade investment)	(15)	
Purchase of fixed assets	(72)	
Sale of patent	5	
Sale of plant and machinery	15	
Received from Customs & Excise on investing activities	8	
Cash flow from investing activities		**(59)**
Financing Activities		
Increase in short-term borrowings	40	
Mortgage of property	25	
Payment of dividend	(5)	
Cash flow from financing activities		**(60)**
Net decrease in cash and cash equivalents		**(2)**
Cash and cash equivalents at 1.1.1990		**26**
Cash and cash equivalents at 31.12.1990		**24**

1/ Example taken from Exposure Draft 54 (July 1990) published by the Accounting Standards Committee, U.K.

Example 2[1/]

This example gives an illustrative cash flow statement for a bank. This statement diverges from the attributions normally required because of the special features of a banking operation.

Cash flow statement for Bank Ltd. for the year ended 31 December 1990

	.... (£'000) ...	
Banking Activities:		
Profit before tax and before extraordinary items	80	
Add items not involving movement of cash:		
- Depreciation charged	10	
- Accruals and prepayments	5	
Cash flow from trading activities		**95**
Increases/(decreases) in deposits	161	
Increases/(decreases) in advances	(152)	
		9
Increases /(decreases) in securities held (other than investment securities)		16
Corporation tax paid		(8)
Cash flow from banking activities		**112**
Investing Activities:		
Purchase of shares in ABC Ltd. (trade investment)	(15)	
Purchase of fixed assets	(105)	
Sale of leasehold building	20	
Cash flow from investing activities		**(100)**
Financing Activities:		
Mortgage of property	25	
Payment of dividend	(5)	
Loan stocks repaid	(25)	
Cash flow from financing activities		**(5)**
Net increase in cash and cash equivalents		**7**
Cash and cash equivalents at 1.1.1990		**26**
Cash and cash equivalents at 31.12.1990		**33**

1/ Example taken from Exposure Draft No. 54 (July 1990) published by the Accounting Standards Committee, U.K.

Value Added Statement

Definition

Value added statements rarely appear in company annual reports and accounts. However, with the growing trend towards increasing disclosure of financial information to employees, the value added statement is likely to become a standard company financial statement. Value added has been defined as the increase in realizable value resulting from an alteration in form, location or availability of a product or service excluding the cost of purchased materials or services.

Structure

The value added statement is basically a rearrangement of figures included in the profit and loss account. The statement shows a firm's sales revenue during a period less bought-in materials and services and how it has been divided among the various stakeholders:

- employees in the form of wages, salaries and other benefits;
- providers of capital in the form of interest to lenders and dividends to shareholders;
- government in the form of taxes and levies.

Depreciation and retained earnings are reinvested to facilitate the expansion of the company. An illustration of value added statement is given below.

Value Added Statement
for the Year ending 31.12.90

		$'000
Sources of Income		
Sales revenue		8,274
Bought-in materials and services		4,742
Value added		3,532
Export subsidy		1,018
Total income		**4,550**
Disposal of Total Income		
To employees		
- Wages, salaries, allowances, etc.	1,490	
To providers of capital		
- Interest on borrowed money	230	
- Dividends to shareholders	120	
To government		
- Company tax	640	
To reinvestment		
- Depreciation	1,200	
- Retained profit	870	**4,550**

Advantages and Disadvantages

The advantages of value added statement are:

(1) it shows clearly the rewards which have been obtained by each interested party: employees, lenders, shareholders, government or the business itself. Where data are provided over a period of years, trends can be established, particularly in terms of value added index, i.e.

$$\frac{\text{employee costs}}{\text{value added}}$$

(2) the value added index can be used as a basis for incentive schemes, which could be a particularly important system in times of inflation and the need for higher productivity;

(3) the employees participate in added value but not in profit;

(4) it can foster a sense of team spirit - cooperative effort of all team members who are involved in adding value.

There are, however, some disadvantages as well:

(1) The meaning is not the same as economists' "Added Value" or that used for value added taxation. This may cause confusion.

(2) Added value, as generally disclosed, will tend to be increased if fixed assets are bought rather than rented and if employees are engaged rather than using sub-contractors.

XII. FINANCIAL REPORTING BY AGRICULTURAL PRODUCERS

There is no set format for financial reporting by agricultural producers. An owner-operated farm has no requirement for financial reporting such as that with which the limited liability companies are required to comply. Farm accounts are normally maintained to satisfy the tax authorities and are mostly based on the cash flows in the farm enterprise. Even within limited companies, financial reporting is not uniform. Specialized enterprises like tea estates have special requirements, for example, in respect of valuation of unsold tea. The following section presents three examples which broadly illustrate the differences in presentation of accounts - the first dealing with an owner-operated farm, the second, a company operated enterprise, and the third, the accounts of a tea estate. It must be noted, in all cases, that the amounts shown in the illustrative financial statements may not necessarily indicate proper customary relationships between accounts.

Accounts of an Owner-Operator

An income statement summarizing the operations of a mixed farm (livestock and crops) is shown below. Although many agricultural producers maintain their accounts on a cash basis, an income statement prepared using the accrual accounting concept is preferred because it will include both cash and non-cash receipts and expenses as well as accounts for inventory changes.

Mixed Farm Enterprise
Income Statement for Year ended 31 December 1990

Revenues		($)
- Livestock sales		160,000
- Crop sales		25,000
		185,000
Operating expenses		
- Hired labour	11,000	
- Livestock	3,000	
- Crops	18,000	
- Fuel	2,000	
- Repairs	5,000	
- Machinery hire	4,000	
- Utilities	1,000	
- Others	300	
Feed purchases	18,000	
Livestock purchases	37,000	
Fixed expenses		
- Land tax	4,000	
- Insurance	2,000	
- Interest	20,000	
- Rent	8,000	
Depreciation		
- Machinery	10,000	
	143,300	
Less Inventory Increases		
- Livestock	3,000	
- Crops and supplies	9,000	
Less value of livestock products consumed on-farm	**1,300**	**130,000**
Net Farm Income		**55,000**

Net farm income (in owner-operated business) is a composite measure incorporating **returns to his/her and his/her family labour, his management** and **his/her contribution of equity in the farm.** All other items have been paid a return for their use in the business. Market prices have been charged for operating and fixed cash expenses; depreciation has been charged on machinery; and interest charged on long-term loan. Only the above three items (family labour, management and owner's net worth in the business) have not been rewarded in the accounts.

Net farm income reflects the monetary value the family had available for living, loan repayment, and for increasing equity in the business or other investment or savings. Part of this income may be reflected in the inventory of livestock or crops, which is a form of investment within the business.

Accounts of the Cattle and Pig Producer Ltd.

The financial accounts of Cattle and Pig Producer Ltd. are given below. The notes to the accounts are representative of the basic types of disclosure that should be made by the farmer; additional disclosures may be appropriate depending on the circumstances. A separate schedule of operating expenses has also been included as supplementary information to support the figures incorporated in the income statement. This schedule is not an integral part of the financial statements, but is considered useful because of the information it provides.

Cattle and Pig Producer Ltd.
Balance Sheet as at 31.12.90

Fixed Assets	**($)**	**Shareholders Equity and Liabilities**	**($)**
Fixed Assets (Note 3)	100,000	Share Capital	96,000
Investment in Feedmill Ltd. (Note 1)	5,000	Profit & Loss account	17,530
Production animals, at cost (Note 2)	15,000	Long-term loan (Note 4)	28,470
Current Assets		**Current Liabilities**	
Inventory		Creditors	5,000
- livestock	65,000	Bank overdraft	51,000
- feed and supplies	12,000	Current portion of long-term loan	4,000
Trade debtors	3,000		
Prepaid expenses	2,000		
	202,000		**202,000**

Cattle and Pig Producer Ltd.
Income Statement for the Year ended 31 December 1990

Revenues	**($)**
Market cattle and pig sale	150,000
Sales of cattle breeding herd	1,000
Crop sales	30
Custom work	700
	151,730
Cost of Animals Sold	
Market cattle and pig	80,000
Cattle breeding herd	800
	80,800
Gross Profit (or Margin)	**70,930**
Operating Expenses	
- Variable (Schedule 1)	39,000
- Fixed (Schedule 1)	16,000
	55,000
Financial Expenses	
- Loan interest	8,000
Other Income	
- Wood & gravel	100
- Gain on sale of fixed assets	500
	600
Income before tax	8,530
Tax	1,000
Net Income for the Year	**7,530**

Net income represents after-tax profit which is available for paying dividends to shareholders. It is assumed that the company has retained the full amount without distribution (unlike the Tea Plantation Ltd. discussed below) to strengthen its capital position.

Notes to Accounts

1. **Significant accounting policies**

Trade debtors for cattle and pigs are based on contracted prices. The company ascertains the collectibility of all accounts at year-end. Any amounts considered doubtful are written off when such a decision is made.

Inventories. Cattle and pig inventories are valued at the lower of their cost and net realisable value. Costs of raising the animals include proportionate costs of breeding, including depreciation of the breeding herd, in addition to the cost of maintaining the animals to maturity. The cost of the purchased cattle and pigs includes purchase cost and the cost of maintaining the animals to maturity.

Feed and supplies inventory is valued at the lower of cost and estimated replacement cost.

Production animals are carried at purchase cost. Amortization is provided on a straight-line basis over the useful lives of the animals, estimated to be 8 years for cattle and 3 years for pigs.

Investment in Feedmill Ltd. represents money invested by the producer in feedmills in order to ensure supplies on a regular basis.

Fixed assets. Fixed assets are carried at cost with depreciation provided on a straight-line basis at the following annual rates:

Land improvement	-	5%
Buildings	-	10%
Equipment	-	20%
Vehicles & farm machinery	-	30%
Silo	-	20%

Costs of maintenance and repairs that do not extend asset life or improve the asset are treated as expenses.

2 **Production animals**

	($)
Cattle breeding herd	14,000
Pig breeding herd	1,000
	15,000

3 **Fixed assets**

	Original Cost	**Accumulated Depreciation**	**Net Cost**
		.... ($)	
Land	63,975	-	63,975
Improvements	2,500	375	2,125
Buildings	30,000	9,000	21,000
Equipment	18,000	10,800	7,200
Vehicles & farm machinery	29,000	26,100	2,900
Silo	7,000	4,200	2,800
	150,475	**50,475**	**100,000**

4. **Long-term loan** - long-term agricultural loan repayable in 8 years at 10% p.a. rate of interest, in equal semi-annual instalments. It is secured by a mortgage on land.

SCHEDULE 1

Cattle and Pig Producer Ltd.
Schedule of Operation Expenses for the Year ended 31.12.90

Variable	**Cattle**	**Pig**	**Total**
		($)	
Custom work	1,800	-	1,800
Drugs & veterinary	630	10	640
Equipment rentals	90	-	90
Feed	17,000	3,000	20,000
Fertilizer	4,000	-	4,000
Gas & oil	2,500	150	2,650
Utilities	700	50	750
Repairs & maintenance	1,200	130	1,330
Seed	900	-	900
Small tools & supplies	850	100	950
Wages and benefits	5,090	800	5,890
	34,760	**4,240**	**39,000**
Fixed			
Depreciation	14,000	125	14,125
Insurance	90	10	100
Management salaries	800	200	1,000
Office expenses	750	25	775
	15,640	**360**	**16,000**
Total Operating Expenses	**50,400**	**4,600**	**55,000**

Accounts of Tea Plantation Ltd.

Tea Plantation Ltd. is a public limited liability company engaged in tea production. Accounts are prepared as in a manufacturing enterprise using accrual accounting concepts. Statements of balance sheet, income and expenditure analysis are set out with supporting explanations.

Tea Plantation Ltd.
Balance Sheet as at 31 December 1990

Fixed Assets		**($)**
Land and improvements		120,000
Plantation costs		150,000
Buildings		60,000
Plant, machinery and fixtures		100,000
Capital work-in-progress		50,000
		480,000
Deferred Assets		
Nursery stocks		20,000
Others		10,000
Current Assets		
Unsold produce	65,000	
Other stock	12,000	
Debtors & prepayments	25,000	
Cash and bank balance	14,000	116,000
Total Assets		**626,000**
Financed by		
Share capital		400,000
Reserves		150,000
Long-term Loans		50,000
Current Liabilities		
Trade and other credits		26,000
Total Capital & Liabilities		**626,000**

Where the cost of land and improvement includes the initial cost of acquisition, the cost of infrastructure development (e.g. roads and bridges) and the cost of initial planting, it is desirable to show the cost of these three items separately in notes to the accounts. The cost of new planting, replanting or major infilling will be shown as capital work-in-progress until plants are mature for harvesting.

The basis of valuation of unsold produce at the year-end must be given. If the net realisable value method is used, but the realisable price is estimated, this fact must be stated. Treatment of replanting expenses should also be explained. When such expenses are charged in the income statement, the amount charged should additionally be shown as a separate item in the income statement.

Plantations normally maintain a nursery to supply plants for replanting and infilling. Nursery expenditure is accounted for separately and a unit cost per plant is arrived at on the basis of successful plant. When above normal mortality occurs, such abnormal loss is not added to the cost of the successful plant. The value of the plants in the nursery is shown in the balance sheet as deferred expenditure and not as a current asset, since most of the plants are used for the creation of a fixed asset.

Tea Plantation Ltd.
Income Statement for Year ended 31 December 1990

	($)
Gross proceeds	256,000
Cost of production	150,000
Working profit	**106,000**
Agency/Head office charges & other expenses	18,000
Net profit from operations	88,000
Other income	2,000
Profit before taxation	90,000
Taxation	30,000
Profit after taxation	60,000
Dividends	10,000
Retained profit for the year	50,000
Balance brought forward	100,000
Balance carried forward	**150,000**

The profit brought forward from the previous year ($100,000) together with after-tax profits of the current year ($60,000), are available for distribution to the shareholders. However, dividends of only $10,000 have been paid out to them.

Gross proceeds are the sale proceeds relating to the total crop harvested during the year, inclusive of the amounts realised on sale of the year's crop after the balance sheet date. The cost of production of the year's harvest is consequently treated as a cost of sales and deducted fully from the gross proceeds to arrive at the working profit.

In plantations which value unsold stocks at cost of production, the gross proceeds would represent the net sale proceeds of the crop sold during the year, which may not necessarily be the total crop harvested for the year. The cost of production for the year, adjusted for the opening and closing stocks valued at cost, will be the cost of sales.

The amount of replanting expenses included in the cost of production or the amount of provision or other charges made for amortization of replanting costs and depreciation of fixed assets is to be disclosed in notes to the accounts.

Analysis of expenditure would vary from plantation to plantation. It is done on a functional basis - plucking, manufacture, land treatment and maintenance, and capital expenditure. Other costs are lumped usually as general charges.

Plucking

Variable costs - wages
- plucking baskets and other materials
- transport of leaf

Fixed cost - supervision

Manufacture

Variable costs - wages
- fuel
- electricity
- packing materials
- transport of made tea

Fixed costs - building repairs and maintenance
- equipment maintenance
- staff of watchers

Land Treatment & Maintenance

- Manuring
- Pests & diseases
- Moss & ferning
- Weeding
- Pruning
- Forking
- Infilling
- Soil conservation

Capital Expenditure

Tea factory expenditure - buildings
- equipment

General assets - roads and bridges
- buildings
- equipment
- vehicles
- housing for workers

General Charges

- Managerial salaries
- Office staff salaries
- Supervisory vehicle costs
- Maintenance of buildings
- Roads maintenance
- Stationery & communications
- Medical benefits & other welfare
- Depreciation

XIII. FINANCIAL RATIO ANALYSIS

The analysis of financial statements is undertaken mainly to evaluate the past performance of an operating entity and forecast its future performance. The previous chapter set out the analysis in terms of absolute dollar values of accounts. This chapter considers how ratios may be more meaningful in assessing a company's situation and performance. This is attempted by first defining, then computing and interpreting some of the important ratios applicable to all entities maintaining accounts on an accruals basis.

Types of Analysis

Financial statements are looked at from various viewpoints by the different parties interested in them. As discussed in Chapter IV, short-term creditors are most interested in the analysis of liquidity or the ability of a project or an enterprise to pay its debts when due. In contrast, long-term lenders are concerned with the earning power and the ability to meet long-term debt requirements. Investors are similarly interested in profitability and efficiency. Management is interested in all those aspects - to repay its debt to both short and long-term creditors as well as earn profits for shareholders. In keeping with the needs of these interested parties, financial ratios can be broadly classified into **liquidity** (or solvency) **ratios**, reflecting the short-term financial position; and **profitability ratios**, reflecting the long-term financial position of an enterprise.

It is important to emphasize that a single ratio in itself is a poor indicator of financial health unless it is supplemented by other ratios and information. In addition, it is necessary to take into account seasonality of operations. Underlying trends may be assessed only through a comparison of raw financial data and ratios referring to the same time of the year or period. For example, it is not proper to compare a December 31 balance sheet with a balance sheet as at June 30. Instead a 31 December balance sheet at a point of time in one year must be compared with the balance sheet for the same month of a different year. Ratio analysis is a rudimentary tool which requires the exercise of judgement and does not provide answers to all of the questions that must be solved.

The profit and loss statement and balance sheets for Tea Ltd., shown below, are used to illustrate the ratios.

Tea Ltd.
Profit and Loss Statement for Year ended 31.12.1989

	$'000
Sales revenue	2,500
Cost of goods sold	1,600
Gross profit	900
Administrative and marketing expenses	200
Operating profit	700
Net non-operating earnings	150
Profit before interest and tax	850
Interest charges	70
Profit before tax	780
Tax	300
Profit after tax	480
Dividends	300
Retained profit	180

Notes:

Production in 1989	=	$2,000,000
Raw material used in 1989	=	$ 800,000
Credit sales	=	$2,000,000
Cash sales	=	$ 500,000
Raw material purchase on credit	=	$ 860,000

Tea Ltd.
Balance Sheet as at 31st December

	1989		1990		1989	1990
Fixed Assets	....		($'000)	....		
At cost					800	1,000
Less accumulated depreciation					200	220
Net Fixed Assets					**600**	**780**
Current Assets						
Raw materials		200		260		
Work in progress		90		120		
Finished stock		100		130		
Trade debtors		800		900		
Short-term investments		-		100		
Cash and Bank balance		80		5		
		1,270		**1,515**		
Less Current Liabilities						
Trade creditors	200		250			
Accrued expenses	10		15			
Proposed dividends	300		300			
Tax payable	60	570	50	615		
Net Current Assets					700	900
Total Assets					**1,300**	**1,680**
Financed by:						
- Ordinary shares (fully paid)					600	600
- Retained profit					200	380
- Long-term loans					500	700
					1,300	**1,680**

Analysis of Short-term Financial Position

Liquidity ratios provide the primary means of judging the short-term financial position of a project or enterprise. Liquidity, defined as the ability to realize assets into cash, has two dimensions: (1) the time required to convert the asset into cash; and (2) the certainty of the realized price. It is usually measured by one or more of the following ratios:

(a) **current ratio;**
(b) **quick asset ratio;**
(c) **stock turnover;**
(d) **debtors' turnover;**
(e) **creditors' turnover.**

Current Ratio

The current ratio is an overall test of liquidity and a rough measure of medium-term solvency because it is heavily dependent on the speed of conversion of stocks into cash. The current ratio is computed by dividing current assets by current liabilities. For Tea Ltd., the ratio for 1990 is:

$$\frac{\text{Current assets at the end of the period}}{\text{Current liabilities at the end of the period}} = \frac{1,515,000}{615,000} = \frac{2.5}{1}$$

Although there are no universally recognized current ratios, a current ratio of 2:1 would generally be considered very good and 1.5:1 would probably be considered the normal acceptable minimum. The more liquid the individual assets, the lower the acceptable ratio. High ratios may be due to poor investment policies, excessive stocks or slow collection of debtors. A low one may indicate a shortage of working capital. Consequently, the components of current assets and current liabilities must be assessed before drawing conclusions on the validity of specific level of ratios. In addition, these must be compared with internal standards or with industry averages. Ratio analysis may have to be supplemented by forecast cash flows for better judgement.

Quick Assets Ratio

The quick assets ratio measures the company's ability to pay creditors in the short-term. Quick assets comprise cash and those debtors, securities and other assets which can be quickly turned into cash with minimal loss. Because the ratio does not involve the valuation of physical assets, it is considered relatively free of measurement problems. This ratio, also known as liquid asset ratio or acid-test ratio, is calculated by dividing quick assets at the end of period by current liabilities at the end of period. For Tea Ltd., the ratio for 1990 would be:

$$= \frac{\text{Quick assets at the end of period}}{\text{Current liabilities at the end of period}} = \frac{1,005,000}{615,000} = \frac{1.6}{1}$$

This is a test of short-term solvency. This ratio is similar to current ratio except that it excludes stocks and work in progress on the grounds that they are invariably the least liquid current assets. As in the case of the current ratio, there are no universal standards for the quick asset ratio. However, a ratio below unity may suggest potential insolvency and the need to obtain further cash resources or overdraft facilities. An overdraft from the bank, shown as a current liability, is in many cases renewed from year to year substituting for longer-term source of funds. Inclusion of this item under current liabilities may in these cases depress the current ratio. A ratio above unity may not necessarily be good if the situation is due to inefficient or unwise use of liquid resources. This ratio may fall in prosperous times because of increased sales which may cause debtors but not cash to rise. Conversely, when business is slowing down, debtors may decrease while cash balances build up.

To the extent that there exists imbalance in the various components of the current assets, the analyst may wish to examine them separately in his assessment of liquidity.

Stock Turnover

Liquidity of stocks is indicated by the stock turnover ratio. This shows the rapidity with which the stocks are turned over into debtors/cash through sales. The ratio is computed by dividing the cost of goods sold by average stock. For Tea Ltd., the ratio is 3.55.

The stock turnover may be expressed in two alternative ways:

1. $\frac{\text{Cost of goods sold}}{\text{Average stock}} = \frac{1,600,000}{450,000} = 3.55$

This tells us the number of times the stock has been turned over in a year to generate sales.

2. $\frac{\text{Average stock x sales days (365)}}{\text{Cost of goods sold}} = \frac{450,000 \text{ x } 365}{1,600,000} = 103 \text{ days}$

This is the inverse of the first ratio and it tells us the period taken to turn over the stock or the average period the stock has remained in the enterprise.

The stock figure used is an average of opening and closing stock for the period. When there are strong seasonal fluctuations, it may be necessary to use an average of monthly or quarterly closing balances. This ratio, like other ratios, must be judged in relation to past and future ratios of the firm compared to similar firms or industry averages or both.

Generally, a high turnover ratio indicates efficient stock management. But a high turnover may be the result of too low a level of stock and frequent stockouts. A low stock turnover ratio may suggest too large an investment in stocks. This may be due to accumulation of obsolete stocks, an imbalance between purchase and usage of raw materials or a build-up of work in progress due to bottlenecks in the production cycle. These can be further investigated by ascertaining individual turnover of finished stocks, raw materials and work in progress as shown below.

$$\text{Finished stock turnover} = \frac{\text{Cost of goods sold}}{\text{Average finished stock}} = \frac{1{,}600{,}000}{115{,}000} = 13.9$$

$$\text{Work-in-progress turnover} = \frac{\text{Annual production}}{\text{Average work in progress}} = \frac{2{,}000{,}000}{105{,}000} = 19.0$$

$$\text{Raw material turnover} = \frac{\text{Annual raw material used}}{\text{Average raw material stock}} = \frac{800{,}000}{230{,}000} = 3.5$$

Conversely a decline in the ratio may be due to a build-up at the year-end of stocks before launching new product line in the following year. This will not indicate an average situation. Another issue in stock management is related to valuation. If the stocks in the accounts are overvalued, then the liquidity of the enterprise is less than what the current ratio alone would suggest.

Debtors' Turnover

Liquidity of debtors is indicated by the debtors' turnover ratio. This shows the rapidity with which the debts are collected. The debtors' turnover ratios may be expressed in two alternative ways:

1. $$\frac{\text{Annual credit sales}}{\text{Average debtors}} = \frac{2{,}000{,}000}{850{,}000} = 2.4$$

   This tells us the number of times the debtors have turned over into cash.

2. $$\frac{\text{Debtors x sales days (365)}}{\text{Annual credit sales}} = \frac{850{,}000 \text{ x } 365}{2{,}000{,}000} = 155 \text{ days}$$

   This is the inverse of the first and it gives the average collection period or the average number of days debtors are outstanding.

When data on credit sales are not available as is the case with published accounts, the total sales figure would be used for the calculation. Where sales are seasonal, an average of the monthly or quarterly closing stock balances would be used instead of the average of annual opening and closing balances.

A decline in this ratio may suggest that the investment in debtors is too large. This may, in turn, be due to debtors taking longer to settle their accounts, a build-up of bad and doubtful debts, poor credit control or simply a change in credit policy, for example by commencing trading on hire purchase terms or by discontinuing cash discounts for prompt settlement of accounts.

Another means of determining the liquidity of debtors is through an age classification of accounts. This method gives more information than the calculation of debtors' turnover because it pinpoints the problem more specifically. The ageing schedule shows distribution of debtors balances according to how long they have been outstanding.

Age of Accounts (days)	**Percent of total value of debtors**
0 - 60	30
61 - 120	10
121 - 180	10
over 180	50

The above example of an ageing schedule shows that the enterprise is having serious collection problems with some of its debtors. Overdues for over six months amount to 50%. Others settle their accounts more promptly thereby bringing down the average collection period to 155 days.

It is important to recognize that both the average collection period and the ageing of accounts are affected by the pattern of sales. Adjustments in analysis should be made. For seasonal sales patterns, it is useful to compare the average collection period and ageing at one point say, 30 June with the average collection period and ageing at the same point (30 June) in another year. The rationale for this is to separate the payment behaviour of customers from changes in pattern of sales. One way to do so is with a conversion matrix of debtors into cash. This shows the amount of credit sales in a given month and the months when these are collected, thereby allowing the analyst to concentrate on the payment pattern of customers month by month as opposed to the combined payments and sales patterns. By tracing the collections in a month to the origin of sales, one can determine if the project's collection experience is improving or deteriorating over time. The principal constraint for this type of analysis is that the required information will have to be obtained from the project or the enterprise since it is not found in the published accounts.

Creditors' Turnover

Creditors' turnover shows the rapidity with which the enterprise pays for supplies received on credit. As in the case of debtors' turnover, this can be measured in two alternative ways:

1. $\frac{\text{Annual credit purchases}}{\text{Average trade creditors at the end of the period}} = \frac{860,000}{125,000} = 6.9$

This shows us the number of times trade creditors are turned over in settlement for supplies received during the year.

2. $\frac{\text{Average trade creditors at end of period x purchase days (365)}}{\text{Annual credit purchases}} =$

$$\frac{125,000 \times 365}{860,000} = 53 \text{ days}$$

This measures the time taken, in days, to pay the trade creditors.

When data on credit purchases are not available, then total purchases can be used for the calculation. Average payment period and ageing of accounts allow the analyst to examine trade creditors in much the same way as trade debtors. Similarly, a conversion matrix for trade creditors is a useful tool for the analysis of situations characterized by either rising or falling sales.

Analysis of Long-term Financial Position

The analysis of the long-term financial position is concerned with measurement of profitability and financial leverage. Important measures of profitability are:

(1) **Gross profit (margin) on sales**
(2) **Net profit (margin) on sales**
(3) **Return on capital employed.**

Measures of leverage are:

(1) **Total debt to total assets**
(2) **Long-term debt to equity**
(3) **Interest coverage**
(4) **Fixed charge coverage.**

Profitability ratios are of two types; those showing profitability against sales generated and those showing profitability in relation to capital employed.

Gross Profit Percentage

Gross profit, which is the difference between sales value and cost of sales, is related to sales to give this ratio:

$$\frac{\text{Gross profit for a period}}{\text{Sales for a period}} \times 100 = \frac{900,0000}{2,500,000} = 36\%$$

Changes in this ratio reflect changes in the relative prices of inputs and outputs, and hence are one basic indicator of the extent to which real exchange rate movements, wage controls, and commercial policy affect company profitability.

This is also a commonly used measure to see whether selling prices are adequate and to determine selling price policy. Mark-up on cost which determines the level of gross profit will vary from one product to another, influenced mainly by the state of competition. A decline in the gross profit percentage would indicate a lower gross profit on sales. This could be due to:

(1) Higher manufacturing costs not passed on in selling prices. This would include not only higher prices paid for raw materials, labour and other services but also higher manufacturing costs due to inefficiencies, e.g. excessive material usage, inefficient workers, stock losses due to pilferage or fixed costs of production being spread over a reduced volume of production.

(2) Reduction in selling prices either due to competition or as an act of policy to stimulate sales volume.

(3) A change in sales product mix, or selling a larger proportion of products with a low margin. In a multi-product enterprise, it is essential to compute gross profit percentage by individual products instead of as an average for all products, to identify the product profitability.

The change in the ratio will not indicate which of the many possible causes has been responsible. This may be revealed by the examination of other ratios.

Net Profit Percentage

Net profit is the difference between sales value and total cost which includes cost of sales, selling, distribution, administration, financial, research and development expenses. The ratio is calculated by dividing net profit by sales as follows:

$$\frac{\text{Net profit before interest and tax}}{\text{Sales for the period}} \times 100 = \frac{850,000}{2,500,000} = 34\%$$

Like the gross profit percentage, this ratio uses sales as the measure to determine profit performance.

Apart from factors mentioned above, a change in the ratio may also be caused by variations in the level of selling, distribution, administrative, research and development expenses as well as the magnitude of net non-operating income. However, both ratios will not correctly reflect profitability, because they do not take into consideration utilization of assets (capital) employed.

Expressed in terms of operating profit, the ratio would be 28%:

$$\frac{700,000}{2,500,000} \times 100$$

Return on Capital Employed (ROCE)

The fundamental measure of profitability is the ratio of profit (return) to capital employed. The term 'capital' and 'profit' are capable of many interpretations and can be expressed at various levels according to purpose of use.

Profit before tax is generally preferred because calculations using profit after tax figures may show trends due simply to changes in the rates of taxation. It may also be noted that the test of efficiency should be based on the gross assets at the disposal of the enterprise irrespective of the method used to finance them. Accordingly, the 'return' means net profit before tax, and capital employed is synonymous with assets employed, and in ratio calculation the usual average would be that of capital employed at successive year ends. But problems of seasonality, new capital introduced or other factors may necessitate taking the average from a greater number of periods within a year. The purpose of obtaining finance in various forms (share capital, long-term loans, short-term finance) is to acquire assets and use them to earn profits.

Concept of Capital Employed 1/	Corresponding Concept of Profit	(ROCE)
1. Share capital + reserves ($890,000)	Profit after interest (short-term and long-term) ($780,000)	88%
2. Share capital + reserves + long term borrowings ($1,490,000)	Profit after short-term interest ($850,000)	57%
3. Share capital + reserves + long-term borrowings + short-term finance (current liability) ($2,148,000)	Profit before interest (short-term and long-term) ($850,000)	39%

1/ Average of 1989 and 1990.

Return on capital employed (Concept 3) measures the productivity of capital and effectiveness of operating management, while return on equity (Concept 1) contains the effects of financial leverage.

An alternative formulation employed indicates the relative contribution of internal and external investment of the enterprise's resources:

Concept of Capital Employed 1/	**Corresponding Concept of Profit**	**(ROCE)**
1. Total or gross assets ($1,490,000)	Profit before interest (short-term and long-term) ($850,000)	57%
2. Operating assets + external investment 3/ ($1,390,000 + ($100,000 = $1,490,000)	Operating profit + investment income ($850,000)	57% 2/
3. Operating assets ($1,390,000)	Operating profit ($700,000)	50%

1/ Average of 1989 and 1990.
2/ Same as (1) because all are operating assets in our example.
3/ Short-term investment - $100,000.

Financial Leverage

Leverage ratios (or gearing ratios) describe the relationship between an enterprise's indebtedness and its equity capital (capital gearing) and the relationship between its interest charges and available earnings (income gearing or interest cover). High gearing indicates a high ratio of debt capital to equity capital and low gearing, the opposite.

Total Debt to Total Assets

The ratio of total debt to total assets (debt ratio) measures the proportion of total funds provided by creditors. Taking the relevant figures from the earlier example for 1990.

$$\text{Debt ratio} = \frac{\text{Total debt}}{\text{Total assets}} = \frac{1,315,000}{1,680,000} = 78\%$$

This index shows the relative importance of debt, both short and long-term in the financial structure of the enterprise. Creditors prefer a low debt ratio as there will be a large cushion (from assets in excess of debts) in the event of liquidation. However, the shareholders usually look for a high debt ratio because, by raising finance through debt, they retain control of the enterprise with limited investment of their own. They will also retain larger earnings if the enterprise earns more profits on the use of borrowed funds than it pays in interest. For example, if assets employed earn 15% and interest on debt only 10%, there is a 5% surplus accruing to the shareholders in addition to earning 15% on their own funds. Leverage, however, cuts both ways; if the return on assets falls to 6%, the differential 4% (10%-6%) must be made up from equity's share of total profits. In the first case, where assets earn more than interest on debt, leverage is favourable; in the second, it is adverse. Thus it may be concluded that the larger this ratio, the more volatile are earnings per unit of equity, the more precarious is the firm's existence and the heavier the drain of interest payments on cash flows.

Debt-Equity Ratio

In addition to the ratio of total debt to total assets, one can compute the following ratio which deals with the long-term capitalization of the enterprise. For the year 1990, the ratio is:

$$\frac{\text{Long-term debt}}{\text{Total capitalization}} = \frac{700{,}000}{1{,}680{,}000} = 42\%$$

(share capital + reserves + long-term debt)

This measure shows the relative importance of long-term debt in the capital structure of the enterprise. An alternative expression of this ratio is:

$$\frac{\text{Long-term debt}}{\text{Equity}} = \frac{700{,}000}{980{,}000} = 71\%$$

(ordinary share capital plus reserves)

Income Gearing or Interest Cover

Interest cover is determined by dividing profit before interest and tax by all interest payable. Because company taxes are computed after interest expense is deducted, the ability to pay current interest is not affected by company taxes. For Tea Ltd. the ratio is 12.

$$\frac{(\ 850{,}000\ )}{(\ 70{,}000\ )}$$

This ratio is used by the creditors/lenders to indicate the vulnerability of the interest payments to a drop in profits. It will reinforce the conclusion based on the debt ratio that the company is or is not likely to face difficulties if it attempts to borrow additional funds.

Cash Flow Coverage

The ability of an enterprise to service debt is related to both interest and principal payments. These payments are not met out of only the earnings. Therefore, a more appropriate coverage ratio should relate to funds internally generated by the enterprise (profit before interest and taxes plus depreciation) to the sum of interest and principal payments. This ratio, known as cash flow coverage ratio, may be expressed as:

$$\frac{\text{Profit before interest and taxes plus depreciation}}{\text{Interest} + \text{principal payments}\left(\frac{1}{1-t}\right)}$$

Where 't' is the company tax rate. As principal payments are made after taxes, it is necessary to adjust this figure by $\left(\frac{1}{1-t}\right)$ so that it corresponds to interest payments which are made before taxes.

Fixed Charge Coverage Ratio

A broader type of analysis would take into consideration not only interest charge but also other fixed charges such as lease payments, preference dividends and possibly even essential capital expenditure, which the enterprise should be able to cover. The ratio is generally known as fixed charge coverage ratio.

$$\text{Fixed charge coverage ratio} = \frac{\text{Profit before interest and taxes}}{\text{Interest charges} + \text{lease payments}}$$

or

$$\frac{\text{Profit before interest and taxes plus depreciation}}{\text{Interest charges} + \text{principal} + \text{lease payments}}$$

Ratios as a System

The ratio of operating profit to operating assets can be further analysed by the pyramid of ratios.

$$\text{Primary ratio} = \frac{\text{Profit before tax}}{\text{Capital employed}} \quad \text{or} \quad \frac{\text{Operating profit before tax}}{\text{Operating assets}}$$

The primary ratio is broken down into secondary ratios which are mathematically linked.

$$\text{Return on capital employed} = \frac{\text{Operating profits before tax}}{\text{Operating assets}} = \frac{\text{Operating profits before tax}}{\text{Sales}} \times \frac{\text{Sales}}{\text{Operating assets}}$$

$$\text{ROCE} = \frac{700{,}000}{1{,}390{,}000} = \frac{700{,}000}{2{,}500{,}000} \times \frac{2{,}500{,}000}{1{,}390{,}000}$$

$$= 50\% = 28\% \times 1.78$$

The ratio of sales value to capital (assets) employed is known as capital (asset) turnover. It indicates the efficiency with which the assets are utilized. The harder the assets are worked, the higher will be the production and capacity utilization and hence sales and profits, which can compensate for a low return on sales thereby yielding a reasonable return on capital employed. Similarly, a high return on sales can offset a low asset turnover and give a reasonable return on capital employed. These possibilities are illustrated as follows:

Return on Capital Employed (%)	=	Asset Turnover (Times)	x	Net Profit (%)
20		10		2
20		20		1
20		2		10
20		1		20

Asset turnover ratio is a better indicator of capacity utilization. At any given book value of assets and set of output prices, the asset turnover ratio will move in direct proportion to the physical volume of sales. So, a comparison of gross profit and asset turnover behaviour can indicate whether changes in gross profit reflect associated relative price shifts, changes in capacity utilization or both.

Secondary ratios can be analysed by further ratios as shown below. This will enable the analyst to spot the problem areas more specifically.

Ratio Pyramid 1/

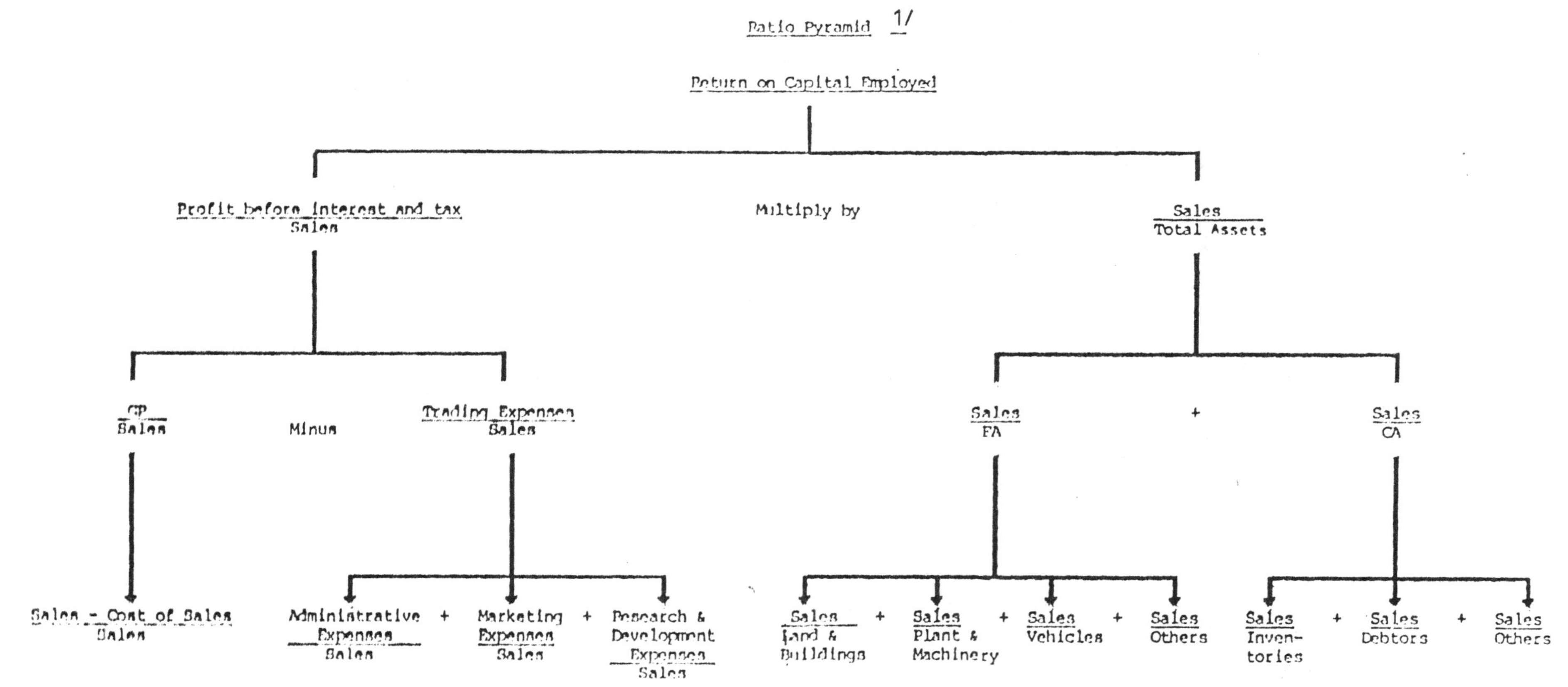

1/ This can be computed for any specific concept of capital employed relating it to the corresponding concept of profit. Presentation here is in terms of the concept of capital defined as total assets and corresponding profit defined as profit before interest and tax.

The object of the above methodical approach is to try and give the answer to why a fall in earnings performance has occurred, and to isolate causes of profit erosion or improvement. For example, the review of the financial statements of a marketing cooperative revealed that, while the return on assets employed was extremely low, the ratio of net profit to sales was high. On further investigation (working through the ratio pyramid) it was revealed that most of the fixed assets (motor vehicles) used for transporting goods for sale were in the workshop being repaired. Hence, while fixed assets had been purchased, they were not being used to generate sales. Thus, while the margin on sales was adequate, not enough sales were being generated to provide an adequate return on the investment in the assets employed. Other cases of a slowing down of fixed asset turnover rate may be due to under-capacity working due to recession, inefficiency in production planning and control or an expansion of plant and facilities on the expectation of increased sales which have not yet materialized. Similarly, a slowing down of the current assets turnover rate signals a slowing down rate of one or more of the components. This may be due to a shortfall in sales but an unchanged volume of working capital; satisfactory sales but more working capital than is necessary (one of the symptoms of undertrading); overstocking; a lengthening of the production cycle, or a longer period of credit given to debtors.

Relevance of Ratios for the Analysis of Public Enterprises

Profitability is probably the main objective of private enterprises and the forces of competition may offer some safeguards to the public which are less evident in public utilities and nationalized industries. The degree of profitability in many public enterprises is determined mainly by the level of selling prices which in turn may be strongly influenced by the policies of government. Assessment of public enterprises' performance in terms of profitability may thus reveal more about government policy than about the enterprises' own efficiency. Furthermore, it ignores the wider economic and social responsibilities of the public enterprises.

Productivity and control of costs are important determinants of profitability. But because pricing policy and consequent profitability (or lack of it) can mask any improvements or reductions in productivity, it is important to use indicators of efficiency and control of costs. The secondary ratios discussed above assume greater importance in the assessment of performance of these enterprises.

Supplementary Ratios

In addition to the ratios discussed above, there are others which may provide further insights into the operation of enterprises in appropriate circumstances. These are based on value-added and employee effort. Some examples are value added per employee, value added per $1 of investment, value added per $1 of sales, sales per employee, profit per employee, and plant and machinery per employee.

Financial Ratios used in the Analysis of the Financial Condition and Performance of Banks

There are a large number of ratios which are used to analyse a bank's financial condition and performance. While these ratios vary somewhat between countries and between banks, their basic objective tends to remain the same, that is, to provide measures of performance in relation to prior years, to budget, and to other banks.

These ratios can be conveniently classified into the following categories:

(1) **Asset quality;**
(2) **Liquidity;**
(3) **Earnings; and**
(4) **Capital adequacy.**

Some of the ratios which the analyst will find useful are given below. Many other more detailed ratios will normally be prepared by the management to review the condition and performance of the bank and its various departments and market segments.

Ratios Measuring Asset Quality - These include:

(a) $\frac{\text{Loan losses (bad debts written off)}}{\text{Total loans}}$

(b) $\frac{\text{Non-performing loans}}{\text{Total loans}}$

(c) $\frac{\text{Provision for loan losses}}{\text{Non-performing loans}}$

(d) $\frac{\text{Profit after tax}}{\text{Loan losses}}$

(e) $\frac{\text{Increase in provision for loan losses}}{\text{Gross income}}$

Liquidity Ratios - These include:

(a) $\frac{\text{Cash and other liquid assets}}{\text{Total assets}}$

(b) $\frac{\text{Inter-bank and money market deposit liabilities}}{\text{Total assets}}$

Earnings Ratios - These include:

(a) $\dfrac{\text{Profit before tax}}{\text{Average total assets}}$

(b) $\dfrac{\text{Profit after tax}}{\text{Average equity}}$

(c) $\dfrac{\text{Interest spread (or margin)}}{\text{Average total assets}}$

(d) $\dfrac{\text{Interest spread}}{\text{Average earning assets}}$

(e) $\dfrac{\text{Interest income}}{\text{Average earning assets}}$

(f) $\dfrac{\text{Interest expense}}{\text{Average interest-bearing liabilities}}$

(g) $\dfrac{\text{Non-interest income}}{\text{Average commitment}}$

(h) $\dfrac{\text{Non-interest income}}{\text{Average total assets}}$

(i) $\dfrac{\text{Non-interest expense}}{\text{Average total assets}}$

Ratios Measuring Capital Adequacy are:

(a) $\dfrac{\text{Equity}}{\text{Total assets}}$

(b) $\dfrac{\text{Equity}}{\text{Risk assets}}$

Main Uses of Ratios

Ratios are useful both internally for management planning and control and externally for making investment and lending decisions. We have so far presented the various financial ratios. Their uses in planning, control and decision-making are now outlined.

A ratio is not meaningful by itself. It must be compared and the value of such comparison depends very much on the validity of the standard against which comparison is made. These standards may be based on present achievements of competitors or industry averages, or targets established in long-term plans or short-term budgets where these are prepared by the enterprise, or historical performance revealed by trend analysis. Trend analysis involves computing ratios of a particular firm for several years and comparing the ratios over time to see if the enterprise is improving or deteriorating. When a trend analysis is combined with comparisons of similar enterprises and the industry averages, the depth of analysis possible is enhanced considerably. A single ratio may only provide a limited view of a situation so that there is often a need to use several ratios in the one context to provide cross reflections and depth to interpretations.

Main Limitations of Ratio Analysis

The main limitation of ratio analysis arises from the inadequacy of information as reported in published accounts. This applies particularly to the definition of profit and valuation of assets. As a result, measures of profitability such as return on capital employed is difficult to compute in a manner which is meaningful for comparison of different companies. For example, two firms may use different depreciation methods or stock valuation methods giving rise to reported profits being lower or higher. Similar differences can occur due to different treatment of provision for bad debts, research and development expenditures. Given this situation, ratios calculated on the basis of published accounts over a period of time are likely to give only an indication of the trend in performance. Even this may be illusory due to distortions created by inflationary conditions. Even if one were able to overcome these shortcomings, the return on capital employed is far from an ideal measure of efficiency because it ignores risk, duration of return, the time element (discounting) and the profitability of discarded alternatives.

Liquidity ratios are subject to less distortions in the valuation of assets. But even here lack of information in published accounts on critical factors such as cash needs of the company and source of cash to meet such needs over the next few months limit the usefulness of these ratios to the particular point in time when the financial statements were prepared. This deficiency is only partially overcome by the historical funds statements which now form part of financial statements of many enterprises.

Finally, the analyst should be careful when judging whether a particular ratio is 'good' or 'bad' and in forming a judgement about an enterprise on the basis of a set of ratios. As we have seen earlier, a high stock turnover may be the result of efficient stock management but it could also be due to a shortage of stock and suggest the likelihood of stockouts.

The project analyst should be able to supplement published information through direct access to the management if he is to make a proper financial evaluation of a project or enterprise by means of ratio analysis.

XIV. ACCOUNTING FOR THE EFFECTS OF CHANGING PRICES

Historical cost accounting is a universally recognized accounting convention which most enterprises continue to use as the preferred method of financial reporting. It is also an accepted basis for the legal accounts of enterprises both in the developed and developing countries. Inflation has seriously tested the reliability of historical cost accounting for reporting operating results and financial positions of companies. There is general consensus on the need to reflect the effects of changing prices in company's financial statements where these are material. In the 1970s approaches to address the issue, based on either the restatement of historical cost information in constant units or the incorporation of **current (or replacement)** costs, were proposed. Nevertheless, neither approach can be said to have gained universal acceptance and there remains disagreement as to the most appropriate method of appraising and reporting the effects of changing prices. This chapter attempts to highlight the major limitations of historical cost accounting to enable the users to understand properly and interpret correctly the financial information presented in such accounts.

Limitations of Historical Cost Accounts

Measurement of assets and liabilities. In historic cost accounts, assets are measured in the balance sheet by reference to their cost subject to a convention that where **value to the business** (in the case of fixed assets) or net realizable value (in the case of current assets) is lower, these figures should be used. The cost is usually the original purchase price or historic cost which is not an up-to-date measure of the resources employed in a business. Many companies try to overcome this deficiency by periodically revaluing some of their fixed assets in their accounts.

Liabilities are entered in the balance sheet at their monetary value so that, for example, a company which has borrowed $10 million will show this in its balance sheet as liability of $10 million, irrespective of when the debt is due to be repaid.

The practice of showing assets and liabilities at cost means that unrealized holding gains are not recognized by the historic cost accounts. Such gains (which arise when an asset is revalued in the accounts at more than its written-down historic cost) are only recognized when assets/liabilities are measured by reference to their value rather than their cost.

Measurement of profit. Historic cost accounting matches current revenues with out-of-date costs. The depreciation charge calculated on the original cost of the fixed asset will be inadequate either to replace the asset consumed during the year or to measure the cost of such consumption. The charge for the cost of stocks consumed will be inadequate to replace them because the stocks are charged at the cost of purchase and not at the cost of replacement.

The effects of holding monetary assets or owing monetary liabilities are ignored in historic cost accounts. If a company holds monetary assets through a period of rising prices, there is a loss involved, represented by the progressive reduction in the goods that can be purchased with the money. Correspondingly, if a company incurs monetary liabilities to be repaid later when prices have risen, the amount borrowed has more purchasing power than the similar amount to be subsequently repaid, resulting in benefits or gains to the company. Because such holding gains or losses attributable to price level changes are not identified, management's effectiveness in achieving operating results may be concealed.

As a result of these deficiencies, reported results may be distorted. The full distribution of profits calculated on that basis may result in the distribution of sums needed to maintain capital. A distribution, which appears well covered when measured against historical cost profit, may appear much less well covered when compared with a measurement of profit that takes account of changing prices.

A consequence of these limitations is that while historical cost information may appear adequate for stewardship purposes, it may provide unsatisfactory indicators for decision-making, either within the company or externally, on matters such as investment and financing decisions, dividend payments, and pricing and pay policies. In practice, many managements have understood the limitations of historical cost information and have devised their own systems of taking account of the effects of changing prices for decision-making purpose. Although details vary, the essential approach is similar, current or replacement costs are substituted for historical cost of inputs.

The limitations of historical cost accounts exist not only in periods of relatively rapid inflation but also when prices are changing more slowly. This is because:

(i) the cumulative effect of low rates of price change over time can be significant: for example, with 5% inflation, prices double about every 14 years;

(ii) the accounting effects of higher previous rates of price change persist over a number of years; and

(iii) specific price changes may be significant even when the rate of general price changes is low.

Remedying the Limitations of Historical Cost Accounting

Some companies have sought to mitigate a number of the limitations of historical cost accounting by adopting a modified historical cost convention, under which certain assets are included in the balance sheets at revalued amounts. However, most companies which use the convention undertake revaluations comparatively infrequently and do not revalue all their assets. The result of this is that many of the limitations still remain.

Some of the limitations of historical cost accounting can be corrected by supplementing it (or replacing it) with information on the effects of changing prices on the results and financial position of the company. Such information can indicate profitability on a basis that makes allowance for the maintenance of operating or real financial capital; it can give a more up-to-date indication of current operating results and management performance; it can indicate more realistic, up-to-date amounts in respect of assets; it can make allowance for the effects of changing prices on monetary items; and it can provide a more realistic indication of trend information. The debate on the appropriate method has often been expressed in terms of a straight choice between two methods - current cost accounting (CCA) in which adjustments are made for specific price changes and constant (or current) purchasing power (CPP) in which adjustments are made for changes in the general level of prices. Although there exist a variety of ways in which a company may enumerate and report the effects of changing prices on its results and financial position, it is appropriate for companies to disclose information about the current year's results and financial position on the basis of: (a) current cost asset valuation using either the **operating or financial capital maintenance concept**; and (b) the nominal currency (e.g. dollar) as the unit of measurement.

Some illustrations. The distortion of reported profits and the effect this can have on the amounts that need to be retained in a business is illustrated by the following examples. A trader in cattle begins the Year 1989 with one cow, which he bought for $20. He has no others, so this is his opening capital. He sells the cow for $30, making a profit of $10 in terms of historical cost. During 1989, general inflation is 10% but the wholesale price of cows increases by 25% to $25 a cow. If he spends his historical profit of $10, he will still have the original money capital of $20 but this will not be enough to replace the cow to begin the next year's trading. To do so without recourse to borrowing, he will need to retain $25 rather than $20; capital is measured in terms of actual operating assets. Under this approach, the trader can withdraw from the business for his own consumption only $5 (i.e. the selling price of $30 less the replacement cost of the cow of $25) without encroaching on the operating assets of the business.

Capital, instead of being viewed in terms of operating assets, can be considered in financial terms. As general inflation in 1989 was 10%, it follows that in order to maintain the **real financial capital** (i.e. the real purchasing power of his original investment), the capital at the end of 1989 needs to be $22, any excess over this amount can be regarded as profit.

These two approaches can be contrasted with historical cost accounting as shown below.

Comparison of historic cost accounting with other systems of accounting

	1	2	3
	Financial Capital Maintenance		**Operating Capital Maintenance**
	Money Capital i.e. Historic Cost Accounts	**Maintenance of General Purchasing Power of Original Investment**	
	.... **($)**		
Sales	30	30	30
Less costs of sales	20 1/	20 1/	25 2/
Operating Profit	**10**	**10**	**5**
Less inflation adjustment to shareholders' funds	-	2	-
Total Gain	10	8	5

1/ Historical cost.
2/ Current cost.

This example shows that if the trader were to distribute or consume the historical cost accounting profit of $10, he would be worse off at the end of 1989 than at the beginning of 1989. This applies whether the method selected for adjusting for inflation is that shown in column 2 or column 3. If he were to distribute or consume $8 indicated in column 2, the general purchasing power of his investment in the business would be maintained but he would still be unable to replace the cow to begin the next year's trading. This demonstrates the different objectives of the two capital maintenance concepts; column 2 is concerned with maintaining the **financial capital in real terms** (i.e. in terms of general purchasing power) whereas column 3 is concerned with maintaining the **operating capital.**

Further examples are given illustrating particular weaknesses of historical cost accounts.

Depreciation. Companies I and U purchased identical fixed assets at the beginning of 1988 and 1989 respectively. Depreciation is assumed to be on a straight-line basis at 20% per year. Price is assumed to have increased by 40% during 1988.

Effect of Changing Prices in the Balance Sheet

	Stable Price end 1989 entries		**40% Price Rise during 1988 end 1989 entries**	
	1 Company I ($)	**2 Company U ($)**	**3 Company I ($)**	**4 Company U ($)**
Fixed assets	1,000	1,000	1,000	1,400
Depreciation	400	200	400	280
Net value	600	800	600	1,120

Columns 1 and 2 show entries in the historic cost balance sheet for the end of 1989 when there has been no change in the price of the asset. Both I and U are assumed to have paid $1,000 for their asset purchased at the beginning of 1988 and 1989 respectively. However, I's asset bought one year earlier than that of U is shown at a net figure after depreciation of $200 less than U. When prices are stable, the entries of $600 and $800 correctly reflect the relative value of I's and U's asset at the end of 1989. Similarly, annual depreciation of $200 reflects the current cost of use of the assets.

Columns 3 and 4 show the entries resulting after a price rise of 40% during 1988, i.e. after I buys the asset but before U has made the purchase. I's balance sheet value is unchanged. However, for U $1,400 is the historic cost when it bought the asset at the beginning of 1989 (due to 40% price rise during 1988). The book value of $1,120 for U's asset indicates the net 'value to the business' of this asset after 1/5 of its useful life has gone, while $600 for U's asset, while still being the written down historic cost of the asset, is a poor approximation to the value of the asset to I. After two years of useful life, a reasonable estimate of the value of the asset to I would be $840 ($1,400 less $560 to allow for passing of two out five years useful life). The correct depreciation charge for the year would be $280 (1/5 x $1,400).

As seen in Chapter XIII, the return on capital employed is a key measure of business performance. As can be observed from the above example, the return on assets stated at historic cost during inflation would result in an inflated figure, both because of a larger profit due to a lower depreciation charge, and due to lower asset values as a result of historic cost written down values. Moreover, performance comparison between companies based on this measure would be misleading.

Stocks

Similar problems arise with stock, although typically to a lesser degree as such assets are held by a company for a much shorter period of time than fixed assets. This is illustrated by an example which compares the historic cost profit and the 'operating profit' of a company trading only in stock, during a year when the price of stock increased by 50%. The company is assumed to have 10 units of stock on hand at the beginning of the year, purchased at a cost of $5 per unit. Fifty units are purchased during the year and the purchase price increases to $10 per unit immediately after the beginning of the year. Selling price is $15 per unit.

Effect of Changing Prices in Profit and Loss Account

	Historic Cost Profit		Operating Profit	
	.... ($)		....	
Sales (50 units)		750		750
Opening stock (10 units)	50		50	
Purchases (50 units)	500		500	
	550		**550**	
Closing stock (10 units)	(100)		(100)	
Historic cost of sales		**450**		**450**
Additional charge to reflect current cost of stock consumed		-		50
Total cost of sales		**450**		**500**
Expenses		250		250
Profit before tax		**50**		**nil**

At the end of the year, 10 units are in hand and the purchase price at that time is still $10 per unit. The historic cost profit and loss account, using First in First out (FIFO) convention, shows a cost of sales figure of $450 after deducting the book value of closing stock (assumed to be $100 since under the FIFO, the units remaining on hand are assumed to be those most recently purchased) from the book value of opening stock plus cost of purchases. Other expenses are assumed to be $250 and sales are assumed to be $750 giving a profit before tax of $50.

The 'operating profit' is of course nil. The current cost of the stock sold (50 units), assumed in this case to be equal to the current purchase price, is $500 not $450. The $50 profit shown in historic cost P and L account is a realized holding gain and is excluded from 'operating profit' by making a charge of $50 equal to the difference between the cost of goods sold (on the FIFO convention) and their value to the business. This charge would be offset by showing the holding gain of $50 separately, elsewhere in the accounts.

Monetary Gain (Loss)

The effect of holding monetary assets and liabilities is illustrated by a simple example. A company has borrowed a sum of $1,000 and held it for a year during which prices have risen by 50%. When repayment is made, the value of the loan to the company would be: \$1,500 ($1{,}000 \times \frac{150}{100}$). But the company would repay only \$1,000 (of course with interest), because the liability is fixed in terms of monetary unit, and it remains constant in monetary terms irrespective of price changes. The company may be said to have made a monetary gain of \$500.

Similarly, a company may suffer a monetary loss when it holds monetary assets. Say, a company has lent \$2,000 to another company for a year during which the price levels have risen by 50%. The company should receive \$3,000 ($2{,}000 \times \frac{150}{100}$) in repayment to be placed in the same position as before (i.e. to enjoy the same purchasing power as before). However, the debtor company would make payment of only \$2,000 as the loan was fixed in monetary value. The company could therefore be said to have incurred a monetary loss of \$1,000.

General and Specific Price Change

In conclusion, an example is given to illustrate the effects of a general and a specific price change on the wealth of an individual.

A rise in price of certain goods from \$1,000 to \$1,400 is a 40% specific change - regardless of what is happening to general prices. However, whether or not the owner of this asset feels much satisfaction over his \$400 appreciation will in great degree depend on how general prices have behaved during the period. Naturally he will prefer his specific rise to outstrip the general rise.

Thus he should use two comparisons to test his progress:

		..\$..
1.	Current price	1,400
	Original price	1,000
	Monetary gain	**400**

and if general prices have risen by say 30%:

		..\$..
2.	Current price	1,400
	Original price restated in current money: ($1{,}000 \times \frac{130}{100}$)	1,300
	Real gain	**100**

In this case, the owner has beaten the general index to make a real gain of \$100.

XV. FURTHER ASPECTS OF ACCOUNTS

This chapter reviews some further aspects of accounts covered in previous chapters. It draws on various pronouncements by both national and international accounting standard setting bodies.

Productive Plants, Trees and Vines

The establishment and growth of perennial plants and trees which are assets for a farm or a company usually take more than one and maybe several years before commercial production levels are achieved. During this period, there are substantial expenditures which need to be incurred for labour, materials and overheads to undertake the various cultivation and maintenance activities. These assets are currently accounted in one of the following ways:

(1) initial outlay capitalized and subsequent expenditures charged directly to operations;

(2) carried as a current asset at market value;

(3) carried as a long-term asset at market value.

The most common practice is the first method, i.e. to charge direct to operations all costs, except initial outlay. Specific segregation and allocation of costs to those assets in the development stage and those in the production stage are recognized to be difficult. As a result, rarely are all the costs incurred up to the productive state capitalized and subsequently depreciated over the useful life of the assets. Good practice demands that all direct and indirect development costs incurred during the development period be capitalized and depreciated over the estimated useful life of the asset.

The cost of the plantings themselves and all material, labour and overheads incurred in establishing and maintaining those assets through the growth state until maturity of the productive state should be included in the cost of the assets.

Since the asset will have a service potential of several years and in order to effect a proper matching of the costs with the revenues generated by the assets, all the costs of acquisition should be matched with all the eventual units of production. Once the assets reach the production stage, they should therefore be depreciated over their estimated productive lives. Considerations in assessing estimated productive life include:

(a) geographical factors such as temperature, humidity and rainfall;

(b) irrigation facilities;

(c) grafting and pruning requirements;

(d) variety and classification of the plant;

(e) plant spacing;

(f) type of rootstock used;

(g) picking or harvesting methods; and

(h) anticipated market demand for product produced.

Productive Livestock

Breeding animals. The cost of breeding young animals should be capitalized and continue to be accumulated until the future use of the animal is decided. If transferred to a breeding herd, the accumulated costs would then be transferred and costs would continue to be accumulated until the breeding process starts (the animal is mature). At that time, the costs will become part of the depreciable cost of the herd and the animals are then depreciated over their estimated productive lives. In the case where the animal is to be sold to another breeder or feeder, the costs would continue to be accumulated until sale. A gain or loss representing the excess or deficiency of sale proceeds over the accumulated costs and expenses of sale would then be recognized. In the case where the animals are to be retained until fattened and sold, the costs of production, care and feeding are accumulated until the date of sale and then charged to cost of sale.

Productive animals. Similar accounting practices are followed as in breeding animals. Their costs should be accumulated and then written off over their estimated useful lives.

Flocks (poultry operations). The accounting principles for flocks are similar to those for production animals. However, the operating cycles are much shorter. The production costs are to be accumulated and depreciated, net of estimated salvage value, over the estimated egg laying period (8-14 months).

Other Tangible Fixed Assets

Fixed assets normally make up a major portion of the total assets of an enterprise and therefore have a significant effect on its financial position. Additionally, the determination of whether an expenditure represents an asset or an expense can have a material effect on the profits of the enterprise. Property, plant, machinery and equipment constitute a high proportion of fixed assets. Asset accounting involves:

(a) ascertaining and recording the acquisition cost of the asset;

(b) depreciating the asset during the period for use; and

(c) considering disposal at the end of the useful economic life of the asset.

Acquisition cost. The cost of an item of property, plant and equipment comprises its purchase price, including import duties and non-refundable purchase taxes, and any directly attributable costs of bringing the asset to working condition for its intended use; any trade discounts and rebates are deducted in arriving at the purchase price. Examples of directly attributable costs are:

(a) site preparation;

(b) initial delivery and handling costs;

(c) installation cost, such as special foundations for plant;

(d) professional fees (architects, engineers, etc.).

Financing costs that are attributable to a construction project and that are incurred up to the completion of construction are sometimes also included in the gross carrying amount of the asset to which they relate. When payment for an item of property, plant and equipment in working condition is deferred beyond normal credit terms, it may be appropriate to capitalize the purchase at the cash price equivalent and to charge the difference between this amount and the total payments as interest over the period of deferral. Administration and other general overhead expenses are not a component of the cost of property, plant and equipment, unless they can be specifically related to the acquisition of the asset or to bringing it to its working condition. Start-up and related preproduction costs would not form part of the cost of property, plant and equipment unless they are necessary to bring the asset to its working condition. An enterprise may decide to charge as an expense an item which could otherwise have been included as property, plant and equipment because the amount of expenditure is immaterial.

In arriving at the gross carrying amount of self-constructed property, plant and equipment, the same principles apply as those described above. Included in the gross carrying amount are costs of construction that relate directly to the specific asset and costs that are attributable to the construction activity in general and can be allocated to the specific asset. Any internal profits are eliminated in arriving at such costs.

Cost inefficiencies in the production of self-constructed assets, whether due to temporarily idle capacity, industrial disputes or other causes, are normally not considered to be suitable for capitalization. It is usually appropriate to have regard to a comparison with the cost of equivalent purchased assets or, if an enterprise makes similar assets for sale in the normal course of business, the cost of producing the assets for sale.

When an asset included in property, plant and equipment is acquired in exchange for shares or other securities in the enterprise, it is usually recorded at its fair value, or the fair value of the securities issued, whichever is the more clearly evident. Subsequent additions to acquisition costs may arise when repairs and overhaul occur and there can be difficulty in determining whether these should be charged against income or added to the gross carrying amount of the asset. Only expenditure that increases the future benefits from the existing asset beyond its previously assessed standard of performance is conventionally included in the gross carrying amount. Examples of these future benefits include:

(a) an extension in the asset's estimated useful life;

(b) an increase in capacity; or

(c) a substantial improvement in the quality of output or a reduction in previously assessed operating costs.

Depreciation and revaluation. The gross carrying amount of depreciable property, plant and equipment is normally recovered on a systematic basis over their useful lives. Depreciation is an integral part of accounting for assets with a finite life. Costs are capitalized with a view to future matching and this matching is achieved when depreciation charges are made. It is not appropriate to omit charging depreciation of a fixed asset on the grounds that its market value is greater than its net book value. This is so because the increase in value is a holding gain whereas the depreciation is an operating charge. If the usefulness of an item or a group of items is permanently impaired, for example by damage or technological obsolescence, the recoverable amount may become less than the net carrying amount. In these circumstances, the net carrying amount is reduced to the recoverable amount and the difference is charged immediately to income.

Sometimes financial statements that are otherwise prepared on a historical cost basis include part or all of property, plant and equipment at a valuation in substitution for historical cost, and depreciation is calculated accordingly. Such financial statements are to be distinguished from financial statements prepared on a basis intended to reflect the effects of changing prices. A commonly accepted method of restating property, plant and equipment is by appraisal, normally undertaken by professionally qualified valuers. Other methods sometimes used are indexation and reference to current prices. Two methods exist for presenting revalued amounts of property, plant and equipment in financial statements. Under one method, both the gross carrying amount and accumulated depreciation are restated in order to give a net carrying amount equal to the net revalued amount. Under the other method, accumulated depreciation is eliminated and the net revalued amount is treated as the new gross carrying amount. The method adopted must be disclosed. In any event, an upward revaluation does not provide a basis for crediting to income the accumulated depreciation existing at the date of revaluation.

Different bases of valuation are sometimes used in the same financial statements to determine the carrying amount of the separate items within each of the categories of property, plant and equipment or for the different categories of property, plant and equipment. In these cases, it is necessary to disclose the gross carrying amounts included on each basis. Selective revaluation of assets can lead to unrepresentative amounts being reported in financial statements. Accordingly, when revaluations do not cover all the assets of a given class, it is appropriate that the selection of assets to be revalued be made on a systematic basis. For example, an enterprise may revalue all its assets on a cyclical basis, or may revalue a whole class of assets within a unit or operating company. Further aspects of depreciation are discussed later in this chapter.

Disposal. An item of property, plant and equipment is eliminated from the financial statements on disposal or when no further benefit to the enterprise is expected from its use and disposal.

Items of property, plant and equipment that have been retired from active use and are held for disposal are stated at the lower of the net carrying amount and net realizable value and are shown separately in the financial statements. Any expected loss is recognized immediately in the income statement. In historical cost financial statements, gains or losses arising on disposal are generally recognized in the income statement. On disposal of a previously revalued item of property, plant and equipment, the difference between net disposal proceeds and the net carrying amount is normally charged or credited to income. The amount standing in revaluation surplus following the retirement or disposal of an asset which related to that asset may be transferred to retained earnings.

Any abnormal losses should be analysed, since they may result from inadequate depreciation rates which showed higher profits in previous periods than were justified, or from poor management policies in the acquisition of assets.

Freehold land is not normally subject to depreciation but circumstances may arise which do require depreciation charges. Changes in the desirability of land because of its access to inputs or markets or for social reasons may result in a decline in value and a need for the book value to be written down. Land is also subject to depletion, when the natural resources it contains are extracted.

Intangible Fixed Assets

Accounting for such assets is almost identical to that for tangibles. First, there is the acquisition when the cost of the asset has to be ascertained and recorded. Secondly, during the period of use, the asset has to be amortized. Measuring the amortization can be a particularly controversial problem with intangibles. Intangible assets are normally amortized equally over their useful lives which would not generally exceed 40 years. Finally, consideration must be given to disposal at the end of the useful economic life of the asset.

Long-term Investments

Trade investment would be disclosed in the balance sheet at cost or market-value whichever is lower. The latter reflects the application of the concept of prudence. The income in the form of dividends arising from the investment would be credited to the profit and loss account. When the market value falls below cost and the fall is judged to be permanent, the difference should be charged to income and the value of the investment would be shown at a reduced figure.

Investments made in associated companies require disclosures as follows:

In the investing company's own accounts:

(a) dividends received up to the accounting date of the investing company; and

(b) dividends receivable in respect of accounting periods ending on or before that date and declared before the accounts of the investing company are approved by the directors.

In the investing group's consolidated accounts:

- The investing group's share of profits less losses of associated companies. Where the investing company has no subsidiaries, it will be necessary for it to adapt its profit and loss account to incorporate the additional information. The amount of investment in associated companies must be disclosed in the consolidated balance sheet in the following way:

(i) at valuation; or

(ii) at the cost of the investments less any amounts written off, and the investing company's (or group's) share of the post-acquisition retained profits and reserves of the associated companies.

An investing company which has no subsidiaries should show its share of its associated companies' post-acquisition retained profits and reserves by way of note to its balance sheet. Investment will be shown in its balance sheet on the cost basis.

Investment in subsidiary companies may involve the investing company in preparing group accounts, in addition to preparing its own accounts.

Stocks and Work in Progress

The determination of profit for an accounting year requires the matching of costs with related revenues. The cost of unsold or unconsumed stocks and work in progress will have been incurred in the expectation of future revenue, and when this does not arise until a later year, it is appropriate to carry forward this cost to be matched with the revenue when it arises. The applicable concept is the matching of cost and revenue in the year in which revenue arises, rather than in the year in which the cost is incurred. If there is no reasonable expectation of sufficient future revenue to cover cost incurred (e.g. as a result of deterioration, obsolescence, or a change in demand), the irrecoverable cost should be charged to revenue in the year under review.

Cost of Stocks and Work in Progress

The cost of stocks and work in progress should comprise all expenditure which has been incurred in the normal course of business in bringing the products or services to their present location and condition. Such costs will include all related production overheads even though these may arise on a time basis. The cost of purchase comprises purchase price including import duties, transport and handling costs and any other directly attributable costs, less trade discounts, rebates and subsidies. Other costs (costs of conversion) comprise:

(a) costs which are specifically attributable to units of production, i.e. direct wages, direct expenses and cost of sub-contracted work;

(b) production overheads, based on the normal level of activity taking one year with another.

In particular, overhead costs of general management should be excluded along with abnormal items such as exceptional spoilage or unusual idle time. It is only appropriate to add in selling overheads when valuing stock for which firm (confirmed) sales contracts exist.

Financing charges including interest would not normally be included, although an exception is the property industry, where for development properties, this is more like a normal production cost. Storage costs would be excluded in most cases since they are not part of the cost of bringing a product to its present location and condition, but merely of keeping it there. Where storage is an integral part of the production process, however, as in the making of whisky, storage costs would be added to cost of stocks and work in progress.

Methods of costing. It is frequently not practicable to relate expenditure to specific units of stocks and work in progress. The basis for arriving at the nearest approximation to cost gives rise to two problems:

(a) the selection of an appropriate method for relating costs to stocks and work in progress (e.g. job costing, batch costing, process costing);

(b) the selection of an appropriate method for calculating the related cost where a number of identical items have been purchased or made at different times (e.g. unit cost, average cost, FIFO, LIFO).

In selecting from the above methods, management must exercise judgement to ensure that the methods chosen provide the fairest practicable approximation to actual cost. Methods such as base stock and LIFO do not usually bear such a relationship. Different bases of stock valuation are illustrated later in this chapter.

Long-Term Contract Work in Progress

The major features of long-term work in progress (e.g a dam construction contract) are that the production or construction time spans several accounting periods, and that in many cases they do not represent standard products so that experience of past costs cannot help forecast the future. Owing to the length of time taken to complete such contracts, to defer taking profit into account until completion may result in the profit and loss account reflecting not so much a fair view of the activity of the company during the year, but rather the results relating to contracts which have been completed by the year end. It is therefore appropriate to take credit for ascertainable profit while contracts are in progress, of course subject to certain limitations. This would mean that the amount at which long-term contract work in progress is stated in the balance sheet should be cost plus any attributable profit, less any foreseeable losses. The progress payments received and receivable would be shown as a deduction therefrom.

Stock Valuation

Different bases of stock valuation will, of course, affect not only the cost of goods sold and in consequence the gross profit and net profit figures, but also the assets side of the balance sheet where the closing stocks as valued appear. The present illustration demonstrates the income effect resulting from using alternative basis of stock valuation - FIFO and LIFO.

The accounts for two successive years are shown, but whereas in the first year the volume of stock at the end of the year is the same as at the beginning, in the second year the volume of stock held is assumed to have increased by the end of that year. The initial stock at the beginning of Year 1 is assumed to have been purchased at $5 per unit. The stock purchased during Year 1 is assumed to cost $10 per unit and the stock purchased during Year 2 is assumed to cost $20 per unit.

Comparison Between FIFO and LIFO

	FIFO		LIFO	
Year 1	($)			
Sales (750 units at $15)		11,250		11,250
Opening stock (50 units at $5)	250		250	
Purchases (750 units at $10)	7,500		7,500	
Closing stock (50 units)	(500)		(250)	
Cost of sales		7,250		7,500
Gross profit		**4,000**		**3,750**
Year 2				
Sales (850 units at $25)		21,250		21,250
Opening stock (50 units)	500		250	
Purchases (900 units at $20)	18,000		18,000	
Closing stock (100 units)	(2,000)		(1,250)	
Cost of sales		16,500		17,000
Gross profit		**4,750**		**4,250**

The FIFO convention assumes that the 750 units sold during the first year consists of the original 50 units purchased at $5 each and 700 of the units purchased at $10 each, a total of $7,250. The LIFO column assumes that the 750 units sold consist of the 750 units purchased at $10 per unit, a total of $7,500. In Year 2 when the volume of stock held is assumed to increase, the same effect is observed. In both cases, the resulting profit figures are different. Moreover, valuation of closing stock appearing in the balance sheets would also differ. Ratios based on profitability and/or asset values will be affected, thus militating against inter-firm comparability.

Another aspect which can give rise to a wide range of possible profit figures is the method of applying the concept of cost or market value basis of valuation. Both terms (cost and market value) are ambiguous. Does market value refer to net realizable value or replacement cost? Furthermore the selected "cost or market" rule could be applied either to each individual item or stock, to particular categories of stock or to the stock as a whole with, of course, a vast range of different stock valuation and net profit figures being possible for any given situation.

Depreciation of Fixed Assets

Assessment of depreciation, and its allocation to accounting periods, involves in the first instance consideration of three factors:

(a) cost (or valuation when an asset has been revalued in the financial statements);

(b) the nature of the asset and the length of its expected useful life to the business, having due regard to the incidence of obsolescence;

(c) estimated residual value.

The cost of the asset is usually determinable with reasonable objectivity but, since estimates of useful life and residual value are highly subjective, so also are depreciation charges (and consequently net profit) and the depreciated value of fixed assets shown in the balance sheet. Discretion is also left on the choice of a depreciation method whether straight line, reducing balance, sum of the years' digits, unit of production, etc.

In recent years, with attention concentrated on the effects of inflation, the charging of depreciation by spreading the historic cost of the asset has been criticized as being inadequate to replace the asset, while also resulting in an overstatement of profits and hence higher taxes being paid. Depreciation performs a dual function:

(1) spreading cost;

(2) retaining resources with a view to replacement.

The mere charging of depreciation will not ensure an actual capacity to replace for two main reasons:

(1) the resources retained require to be in liquid form for replacements to be purchased;

(2) the amount of resources retained by way of depreciation charges based on historic cost over the life of the old asset will total at most to the past acquisition cost, which in times of inflation will be inadequate to finance the replacement at higher prices.

To overcome this problem, some advocate that the excess of any amounts required for replacement over and above what is provided by depreciation on historic cost basis should be provided for by appropriating profits to a "Replacement Reserve", which is in fact quite commonly done. The validity of this procedure is questionable because:

(a) this does not avoid the overstatement of net profit in the profit and loss account nor the levying of tax on that overstated profit;

(b) it is essentially a means of classifying retained profits rather than making a charge in calculating those profits.

The choice of a suitable basis for determining annual depreciation charges can be made from the following alternatives.

(1) historical cost basis - spreading cost over assets' useful life;

(2) current replacement cost basis without adjustment for under-provision of previous years, which will match current costs with current revenues but will not provide enough funds to replace the asset;

(3) current replacement cost basis plus adjustments for under-provision of previous years but this will not match current cost against current revenue.

As in most areas of accounting controversy, choice among these alternatives can only be made by reference to the objectives sought to be achieved by financial accounting reports.

If the objective is to satisfy the stewardship function of the directors, then historic cost basis may be suitable. If the objective is to arrive at a meaningful figure of operating profit the unadjusted current cost basis may be supported. If the objective is to ensure firms' capacity to replace the assets, then an adjusted current basis may be considered.

Extraordinary Gains and Charges (or Losses)

Extraordinary items derive from events outside the ordinary activities of the business; they do not include items of abnormal size and incidence which derive from the ordinary activities of the business.

There are currently two different viewpoints on the treatment of extraordinary items. One view is that in order to avoid distortion, the profit and loss account for the year should include only the normal recurring activities of the business. It should therefore exclude the extraordinary items, which should be taken direct to reserves. The other view is that the profit and loss account for the year should include and show separately all extraordinary items which are recognized in that year. The latter view of "all-inclusive" profit is preferable for the following reasons:

(a) inclusion and disclosure of extraordinary items will enable the profit and loss for the year to give a better view of a company's profitability and progress;

(b) exclusion, being a matter of subjective judgement, could lead to variations and a loss of comparability between reported results of companies; and

(c) exclusion could result in extraordinary items being overlooked in any consideration of results over a series of years.

The classification of items as extraordinary will depend on the particular circumstances - what is extraordinary in one business will not necessarily be extraordinary in another. Subject to this, examples of extraordinary items could be the profits or losses arising from the following:

(a) the discontinuance of a significant part of a business;

(b) the sale of an investment not acquired with the intention of sale;

(c) writing off intangibles, including goodwill, because of unusual events or developments during the period; and

(d) the expropriation of assets.

Items which, though abnormal in size and incidence, are not extraordinary because they derive from the ordinary activities of the business, would include:

(a) abnormal charge for bad debts and write-offs of inventory and research and development expenditure; and

(b) abnormal provisions for losses on long-term contracts.

Such items are normally termed exceptional items, and are explained in a note to the accounts.

Research and Development Expenditure

The accounting treatment of research and development expenditures is governed by the concepts of accruals and prudence. Expenditure incurred on pure and applied research can be regarded as part of a continuing operation required to maintain a company's business and its competitive position. In general, one particular period rather than another will not be expected to benefit and therefore it is appropriate that these costs should be charged to profit and loss account as they are incurred.

Expenditure on development of new and improved products is normally undertaken with a reasonable expectation of specific commercial success and of future benefits arising from the work, either from increased revenue or from reduced costs. On these grounds, it may be argued that such expenditure should be deferred to be matched against the future revenue. This is valid in the following (restrictive) circumstances:

(a) there is a clearly defined project;

(b) the related expenditure is separately identifiable; and

(c) the outcome of such a project has been assessed with reasonable certainty as to:

 (i) its technical feasibility;

 (ii) its ultimate commercial viability considered in the light of factors such as likely market conditions;

 (iii) if further development costs are to be incurred on the same project, the aggregate of such costs together with related production, selling and administration costs are reasonably expected to be more than covered by related future revenues; and

 (iv) adequate resources exist or are reasonably expected to be available to enable the project to be completed and to provide any consequential increases in working capital.

Where these conditions do not exist, development expenditure should be written off in the year of expenditure.

XVI. FINANCIAL ANALYSIS OF LENDING INSTITUTIONS 1/

The discussion so far has focused on the financial impact of project investments on participants such as agricultural producers and enterprises. This chapter is devoted to the analysis of the banks involved in project implementation. For the purpose of this discussion, the term 'bank' includes all financial institutions, one of whose principal activities is to take deposits and borrow with the objective of lending and investing for gain. The analysis is based on a review of historical and forecast financial statements, and concentrates on the main performance indicators - **earnings performance, liquidity, capital adequacy** and **asset quality.**

As a starting point, financial statements will have to be summarized as proposed in **suggested outlines** to allow meaningful analysis of financial performance. The outlines of balance sheet, income statement and cash flow statement are presented at the end of the chapter. The level of financial analysis will vary depending on the local legislative and regulatory framework, availability and adequacy of data, and the quality of audit. However, certain basic requirements must be met before any reasonable conclusions can be drawn on the concerned bank's financial performance. These are discussed in the following paragraphs.

Earnings Performance

It is recognized that a bank need an adequate level of profits, firstly to provide an appropriate return to the shareholders; secondly, to reassure depositors that the business is sound and competently managed; and finally, to maintain and expand its capital base, in order to satisfy prudential criteria and facilitate business growth. Most of the profits earned by the bank derive from using its own funds to earn interest and by borrowing on more favourable terms than those on which it lends. The overall objective is to achieve a reasonable **margin** between the rate that it pays for its funds and the amount that it earns by lending or investing those funds.

Commercial enterprises use the concept of **margin** to denote the difference between the cost of sales and the selling price. Banks act in a similar manner, but deal in currency and not goods. The banks acquire funds - on current account, savings, time deposits, certificates of deposit etc. promising to pay interest, and invest these funds in assets - loans, investments etc., - for which they receive interest and fees. Computation of the spread is illustrated below.

1/ This chapter draws heavily on the Annex A of the World Bank's draft Operational Directive on Financial Sector Operations.

Interest cost	1990 (%)
Average cost of new borrowings	7.7
Average cost of outstanding loans and deposits	7.4
Average cost of total funds (equity + borrowings + deposits + loans)	6.3
Interest income	
Average returns on loans disbursed and outstanding	7.9
Average returns on liquid investments	8.2
Average returns on earning assets	8.3
Profitability	
Interest spread between return on earning assets & cost of total funds	2.0 (8.3-6.3)

The **interest margin** (also known as **intermediation margin** or **interest spread**) together with other income such as commission, investment income, and fees for services should generate a surplus after meeting the operating expenses. The resulting surplus, which is the profit after tax, should be adequate in real terms to provide a positive return on equity. The adequacy of the level of earnings is evaluated using the following indicators:

(a) **interest income as % of average total assets (ATA);**

(b) **net interest income as % of ATA;**

(c) **net interest income after provision for losses as % of ATA;**

(d) **interest spread on loans (average % interest income on loans minus average % cost of borrowed funds);**

(e) **operating expenses as % of ATA;**

(f) **profit after tax as % of (i) ATA and (ii) equity at the start of the year;** and

(g) **debt service cover ratio.** (The debt service cover ratio is generally relevant only to development finance institutions and a suggested outline for its calculation is given at the end of the chapter.)

Lending risks. There are a number of different types of lending 'risks' that may affect the interest margin charged to a borrower - **credit risk, liquidity risk, interest rate risk** and **currency risk.** The most significant risk in a bank is usually **credit risk,** which is the risk of being unable to collect amounts due from customers and counterparties due to factors such as deterioration in their financial condition or unwillingness to pay. This may be partially covered by charging a higher interest rate on loans to borrowers whom the bank considers present a higher risk. Apart from the characteristics of individual borrowers, two major criteria that a bank will bear in mind when assessing credit risk are country and industry risk. Normally limits are established by the bank for the proportion of its total portfolio that is committed to any one industry, commodity or if the bank operates internationally, country.

A bank is able to **'borrow short'** and **'lend long'** (known in the industry as **maturity transformation**) because of the size and spread of its resources. As a result, it faces the **liquidity risk** arising from the possibility of the bank having insufficient funds to meet the demands of depositors. To cover this liquidity risk, a bank must maintain a prudent level of liquid assets and may also arrange standby facilities to borrow in times of emergencies. **Interest rate risk** arises from the fact that some of a bank's borrowing and lending will either be for a very short term, needing to be regularly renewed at market rates, or be tied to a variable rate of interest. The risk of loss can arise from mismatches between interest rates on deposits and loans (arising due to changes in rates in one not compensated by similar changes in the other), and decreases in market values of securities due to increases in interest rates. The bank's main **currency risks** arise from not matching its currency assets and liabilities, and from exchange control.

Liquidity

A bank's liquidity is its ability to meet deposit withdrawals, maturing liabilities and loan requests without delay. The mix of these obligations and their incidence in any period of time will vary between different banks, and this makes a single measure or ratio invalid. Therefore, it is necessary to look beyond liquid assets and take into account the bank's access to external funds in case of a liquidity crisis (e.g. central bank facilities, funds purchased from the money market). Nevertheless, the following ratios are used as partial indicators of the effectiveness of liquidity management of a bank.

(a) **'Hot money' ratio.** This ratio of basic market-sensitive liabilities to loans and investments illustrates a bank's reliance on purchased funds; and

(b) **Loan-to-deposit ratio.** This is expressed as total outstanding loans as % of total deposits. Usually a maximum ratio of 70% is considered a prudent limit but final judgement should take into account the country conditions.

For DFIs, liquidity analysis which differs from that described for banks, requires a review of: (a) the relationship between new commitments for loans and investments to the available resources; and (b) liquidity maintained to meet maturing obligations - operating expenses and debt servicing obligations.

Capital Adequacy

A bank must have sufficient capital to absorb risk of losses inherent in the assets of the business so as to protect depositors and other creditors. For DFIs with no or limited access to deposits and liquidity, capital must be adequate to satisfy its maturing obligations in the event of a material fall in its loan collections. Prescription of a precise numerical guideline for the capital needs of all institutions or for groups of institutions would be inappropriately inflexible. Such an approach would endorse overtrading by some and be harmfully restrictive to others. However, it is generally believed that commercial banks should maintain a minimum capital adequacy ratio of 8%, the established DFIs at least 10%, but new DFIs and those with not very satisfactory performance at least 20%. A suggested outline of capital adequacy analysis is given at the end of the chapter.

There are two methods of assessing the adequacy of capital - **'free resources' ratio** and a **'risk asset' ratio.** The first broadly relates current liabilities to capital resources, excluding that part of capital which is invested in infrastructure and other non-banking assets. The second relates the risks of losses which are inherent in the bank's total portfolio to the capital which is available to finance such losses. The two most important objectives of capital ratios are: (i) to ensure that the capital position of the bank is regarded as acceptable by its depositors and other creditors; and (ii) to test the adequacy of capital in relation to the risk of losses which may be sustained. The risk asset ratio is considered superior in judging the adequacy of capital and it is measured in two ways.

The first relates equity capital to total assets - $\frac{\text{equity}}{\text{total assets}}$

which indicates the percentage decline in total assets that could be covered with equity. A principal weakness of this ratio is that it fails to recognize that some assets such as cash and government bonds or government bills involve no default or market risk. As a result, a second ratio was developed relating equity capital to risk assets. This ratio is properly called a risk asset ratio - $\frac{\text{equity}}{\text{risk assets}}$.

The traditional way of relating a bank's business to its capital has been by the **gearing ratio**. This relates a bank's capital to the total public liabilities of the bank, which are generally defined to include customers' deposits and other creditors excluding contingent liabilities, less any intra-group assets. The risk asset ratio relates the minimum desired level of capital to the different categories of assets weighted by their respective degrees of risk ratio. In some countries, e.g. the United States of America and the United Kingdom, off-balance sheet risks are also to be included in this system. A suggested outline for risk weighing of assets is given below.

Suggested outline for weighting of risks in a bank's assets

Risk Classification	Asset Composition	% of Asset Values Recommended as Capital Requirement
a) Without Risk Assets	Cash on hand, bank balances, selected marketable foreign & domestic short-term securities, other assets of comparable quality and short-term maturity.	None
b) Low Risk Assets	Loans and investments that have less than normal credit risks and that may be readily pledged or sold, including government securities with maturities over three years, loans guaranteed by government, loans secured by similar assets (or by savings passbooks, cash value of life insurance, negotiable securities, prime commercial paper, etc.), and buildings having readily convertible sales or lease value.	5% - 10%
c) Moderate Risk	Assets (including loans) with normal or slightly above normal risks such as adverse economic conditions, uncovered foreign exchange risk assumed by borrowers, over-extended loan disbursements, insufficiency of security guarantee, and unfavourable record of obligator.	10% - 25%
d) High Risk Assets	Loans and stocks classified as doubtful by appropriate Bank Officers, or external examiners, defaulted securities and real estate that may not legally be acquired except by foreclosure.	50% (depending on inflation rate)
e) Fixed Assets, Furniture & Equipment	Mixed assets without ready resale or market value, or which may not be disposed of without having negative effect on bank's operations.	100%

The creation of the risk asset ratio system reflects several factors. A major one is the desire of the authorities to catch up with the innovations in banking technique and in the securities markets in recent years, so as to ensure that the capital of the banks is in an appropriate relationship to the types of risks which banks face. Second, there is a general perception that the banks may have too low a level of capital in relation to their risks, and perhaps some of the assets in their balance sheets are overvalued. Third, the risk asset ratio system has been discussed internationally among supervisors notably at the Basle Committee, and has become the norm in relation to international banking.

In defending itself against loss, the bank has both its capital and its continuing profits. Since banks expect to make profits on a continuing basis, the expected level of loss is covered, in a healthy organization, by the continuing profits. Thus, the function of capital is not to cover the average loss but to cover the peak losses, concentration in assets or a high co-variance of loss. However, the risk is based on the portfolio structure of the assets, whereas the risk asset ratios are calculated on the individual components of the balance sheet. At present, there is no specific link of risk asset ratio policy to the concept of portfolio structure. No work appears to have been done on the portfolio and capital structure of banks to calculate what risk asset ratios are needed for safety. Given the very powerful effects of these ratios on bank policy, on competition between banks, and on competition between different financial centres, it is essential that these ratios be set at the appropriate level.

Asset Quality

The determination of asset quality is the most critical part of the bank's financial analysis, and requires minimum uniform supplementary data usually not provided in the published financial statements. The main points to be reviewed are given below.

Portfolio classification and provision for loan losses. This analysis classifies risk assets (loans, advances, overdrafts, guarantees, investments, etc.) by performance (e.g. repayment performance and financial position of the borrower) to provide the basis for estimating provisions for possible losses. Classification guidelines vary from country to country, but the most common practice is to divide the portfolio in the following four categories:

(i) **Normal risk assets** which are problem-free and are performing according to the agreed amortization schedule.

(ii) **Sub-standard assets** performing satisfactorily at present, but certain indicators suggest that some potential difficulties may arise (e.g. losses on recent operations, deficiency in documents).

(iii) **Doubtful assets,** in respect of which a non-performance has occurred; and

(iv) **Loss** in which all or a part of the loan is not expected to be recovered.

For each of the above categories, provisions for possible losses are created representing an estimate of losses. It is necessary to review the provisioning guidelines and decide if they represent a realistic estimate of the potential portfolio losses.

Overdraft financing analysis. Overdraft financing, whereby banks let customers overdraw their current accounts, is a common form of financing in many developing countries. Evaluating such loans in the absence of strong regulatory rules and system is difficult. Generally, an overdraft account should be considered non-performing and appropriate provisions made if:

(i) the account has been inactive for 180 days or more (i.e. either there has been no deposit or deposits have been less than the interest accrued for that period);

(ii) the outstanding balance exceeds by more than 10% the authorized limit for 60 days or more;

(iii) there is an overdraft outstanding for 90 days without any authorized limit; and

(iv) the authorized limit has been cancelled or the customer is under foreclosure proceedings.

Single borrower exposure. Prudential regulations normally provide for a bank's maximum exposure limits for individual customers, both in terms of kinds of exposure (e.g. loans, equity investment, guarantee, etc.) and total exposure. DFIs, if not required by local regulations, should set such prudential guidelines. Such exposure limits apply to net outstanding commitments (i.e. amount committed minus repaid and/or cancelled) for all types of financial assistance including off balance sheet items. In countries lacking adequate prudential guidelines and policies, and their enforcement, the portfolio is analysed in terms of concentration of portfolio in a few enterprises or groups (including those affiliated with bank's directors), a particular economic sector, a particular country or region to determine where excess exposure requires corrective measures.

Arrears analysis. The most important indicator for portfolio infection is the proportion of the portfolio affected by arrears as a percentage of total portfolio. Arrears analysis can take several forms. This analysis should be done for each performance category of the portfolio, using ageing of debtors. Arrears analysis is required for all DFIs and for the term loan portfolio (maturity of one year or more) of all banks. Past dues, overdues and arrears are generally used interchangeably.

Collection analysis. It is necessary to determine how much of the amounts due are actually collected in cash in a year. Banks should, at a minimum, demonstrate a collection rate which - given its lending margin - avoids erosion of capital. For this purpose, the following three ratios are calculated using the suggested outline given below.

(i) Arrears collection rate (total principal and interest amount collected against arrears at the start of the year as a percentage of total arrears at the start of the year);

(ii) Current collection rate (total amount of principal and interest collected against amounts falling due in the year as a percentage of total amounts due in the year); and

(iii) Total collection rate (total amount of principal and interest collected during the year as a percentage of total of arrears at the start of the year and falling due during the year).

The total of ratios (i) and (ii) should equal (iii).

Suggested outline for calculations of collection ratio (or recovery rate)
Collection ratio for the period 1988-1990 (in $ million)

Item	1988		1989		1990	
	Principal	Interest	Principal	Interest	Principal	Interest
1. Arrears at the beginning of the year						
2. New amounts falling due during the year						
3. Total to be collected (1 + 2)						
4. Cash collections against: (a) Arrears (b) Current dues						
5. Total collections (4a + 4b)						
6. Arrears at end of year (3-5)						
7. Amounts rescheduled during year						
8. Net arrears at end of year (6-7)						
Ratios						
Arrears collection rate = 4a ÷ 1						
Current collection rate = 4b ÷ 2						
Total collection rate = 5 ÷ 3						

Rescheduling standards. Rescheduling is necessary from time to time, but should be done within explicit and realistic policies. Loans are rescheduled only when: (i) the customers are sound and creditworthy; and (ii) the bank before rescheduling, has satisfied itself that all necessary steps have been taken to enhance the repayment capacity of the borrower. Rescheduling should not be used for window dressing the portfolios and earnings. In general, not more than 20% of a bank's portfolio should have been rescheduled, and not more than 5% of the portfolio should have been rescheduled more than once.

Financial Projections

The analyst will generally require financial projections for participating banks to demonstrate their viability and creditworthiness. For banks undergoing major institutional and financial restructuring, the financial projections are necessary to support the validity of the restructuring programme. Finally, for DFIs, projections help identify its external financing needs. The following guidelines should be considered in either assisting the bank in the preparation or reviewing the financial projections.

(a) **Period covered.** Financial projections cover at least three fiscal years after the expected date of approval of the project loan or the period of commitment of the loan, whichever is longer.

(b) **Scope and coverage.** The projections comprise: (i) the basic financial statements (balance sheet, income statement, and funds flow statement) in spreadsheet format supplemented by data on projected operational programme (commitments, disbursements, collection/sales, and outstanding amounts of various instruments of financial operations and magnitude of other forms of products and services). The projections are based on the current prices and also include critical indicators of financial position and performance.

(c) **Underlying assumptions.** The analyst should carefully review the assumptions underlying the financial projections, particularly: (i) costs of borrowings; (ii) interest and dividend income; (iii) collection rates; (iv) provisions for possible losses; (v) administrative and overhead expenses; and (vi) taxation on income.

(d) **Sensitivity tests.** To examine the effect of certain critical cost and income variables on the projected financial position of the bank, certain sensitivity tests are applied. The most appropriate variables for such tests would be the level of volume of operations, intermediation margins and collection rates.

The review should highlight major trends, assumptions, and an overall assessment of the future financial position and creditworthiness of the bank. The emerging issues are also noted along with remedial measures which are proposed to be taken.

Key Elements of Banks' Financial Statements

This section is based largely on the recommendations of the International Accounting Standards Committee, which are set out in the International Accounting Standard 30: "Disclosures in the Financial Statements of Banks and Similar Financial Institutions" approved in June 1990 for publication in August 1990.

Income statement. A bank should present an income statement which groups income and expenses by nature and discloses the amounts of the principal types of income and expenses. Generally, the disclosures in the income statement or the notes to the financial statements should include, but are not limited to, the following items of income and expenses:

- Interest and similar income
- Interest expenses and similar charges
- Dividend income
- Fee and commission income
- Fee and commission expense
- Gains less losses arising from dealing securities
- Gains less losses arising from investment securities
- Gains less losses arising from dealing in foreign currencies
- Other operating income

- Losses on loans and advances
- General administrative expenses
- Other operating expenses

Balance sheet. A bank should present a balance sheet that groups assets and liabilities by nature and lists them in an order that reflects their relative liquidity. The disclosures in the balance sheet or the notes to the financial statements should include, but are not limited to, the following assets and liabilities:

Assets
- Cash and balances with the central bank
- Treasury bills and other bills eligible for rediscounting with the central bank
- Government and other securities held for dealing purposes
- Placements with, and loans and advances to, other banks
- Other money market placements
- Loans and advances to customers
- Investment securities

Liabilities
- Deposits from other banks
- Other money market deposits
- Amounts owed to other depositors
- Certificates of deposits
- Promissory notes and other liabilities evidenced by paper
- Other borrowed funds.

Maturities of assets and liabilities. The matching and controlled mismatching of the maturities and interest rates of assets and liabilities is fundamental to the management of a bank. It is unusual for banks ever to be completely matched since business transacted is often of uncertain term and of different types. An unmatched position potentially enhances profitability but can also increase the risk of losses.

The maturities of assets and liabilities and the ability to replace, at an acceptable cost, interest-bearing liabilities as they mature, are important factors in assessing the liquidity of a bank and its exposure to changes in interest rates and exchange rates. In order to provide information that is relevant for the assessment of its liquidity, a bank discloses, as a minimum, an analysis of assets and liabilities into relevant maturity groupings.

The maturity groupings applied to individual assets and liabilities differ between banks and in their appropriateness to particular assets and liabilities. Examples of periods used include the following:

- up to 1 month
- from 1 month to 3 months
- from 3 months to 1 year
- from 1 year to 5 years
- from 5 years and over.

Frequently the periods are combined, for example, in the case of loans and advances, by grouping those under one year and those over one year. When repayment is spread over a period of time, each instalment is allocated to the period in which it is contractually agreed to be paid or received.

It is essential that the maturity periods adopted by a bank are the same for assets and liabilities. This makes clear the extent to which the maturities are matched and the consequent dependence of the bank on other sources of liquidity. Maturities may be expressed in terms of: (a) the remaining period to the repayment date; (b) the original period to the repayment date; or (c) the remaining period to the next date at which interest rates may be changed.

The analysis of assets and liabilities by their remaining periods to the repayment dates provides the best basis to evaluate the liquidity of a bank. A bank may also disclose repayment maturities based on the original period to the repayment date in order to provide information about its funding and business strategy. In addition, a bank may disclose maturity groupings based on the remaining period to the next date at which interest rates may be changed in order to demonstrate its exposure to interest rate risks. Management may also provide, in its commentary on the financial statements, information about interest rate exposure and about the way it manages and controls such exposures.

Repayment maturities may be expressed in terms of the remaining period to either the contractual maturity date or the effective maturity date. In many countries, deposits made with a bank may be withdrawn on demand and advances given by a bank may be repayable on demand. However, in practice, these deposits and advances are often maintained for long periods without withdrawal or repayment; hence, the effective date of repayment is later than the contractual date. Nevertheless, a bank discloses an analysis expressed in terms of contractual maturities even though the contractual repayment period is often not the effective period because contractual dates reflect the liquidity risks attaching to the bank's assets and liabilities.

Some assets of a bank do not have a contractual maturity date. The period in which these assets are assumed to mature is usually taken as the expected date on which the assets will be realized.

The users' evaluation of the liquidity of a bank from its disclosure of maturity groupings is made in the context of local banking practices, including the availability of funds to banks. In some countries, short-term funds are available, in the normal course of business, from the money market or, in an emergency, from the central bank. In other countries, this is not the case.

In order to provide users with a full understanding of the maturity groupings, the disclosures in the financial statements may need to be supplemented by information as to the likelihood of repayment within the remaining period. Hence, management may provide, in its commentary on the financial statements, information about the effective periods and about the way it manages and controls the risks and exposures associated with different maturity and interest rate profiles.

Concentration of assets, liabilities and off balance sheet items. A bank discloses significant concentrations in the distribution of its assets and in the source of its liabilities because it is a useful indication of the potential risks inherent in the realization of the assets and the funds available to the bank. Such disclosures are made in terms of geographical areas, customer or industry groups or other concentrations of risk which are appropriate in the circumstances of the bank. A similar analysis and explanation of off balance sheet items is also important. Geographical areas may comprise individual countries or regions within a country; customer disclosures may deal with sectors such as governments, public authorities, and commercial and business enterprises.

The disclosure of significant net foreign currency exposures is also a useful indication of the risk of losses arising from changes in exchange rates.

Losses on loans and advances. It is inevitable that in the ordinary course of business, banks suffer losses on loans, advances and other credit facilities as a result of their becoming partly or wholly uncollectable. The amount of losses which been specifically identified is recognized as an expense and charged against income and deducted from the carrying amount of the appropriate category of loans and advances as a provision for losses on loans and advances. The amount of potential losses not specifically identified but which experience indicates are present in the portfolio of loans and advances is also recognized as an expense and charged against income and deducted from the total carrying amount of loans and advances as a provision for losses on loans and advances. The assessment of these losses depends on the judgement of management; it is essential, however, that management applies its assessments in a consistent manner from period to period.

Local circumstances or legislation may require or allow a bank to make charges against income for losses on loans and advances in addition to those losses which have been specifically identified and those potential losses which experience indicates are present in the portfolio of loans and advances. Any such charges represent appropriations of retained earnings and not expenses in determining net income for the period. Similarly, any credits resulting from the reduction of such charges result in an increase in retained earnings and are not included in the determination of net income.

Users of the financial statements of a bank need to know the impact that losses on loans and advances have had on the financial position and performance of the bank; this helps them judge the effectiveness with which the bank has employed its resources. Therefore, a bank discloses the aggregate amount of the provision for losses on loans and advances at the balance sheet date and the movements in the provision during the period. The movements in the provision, including the amounts previously written off that have been recovered during the period, are shown separately.

A bank may decide not to accrue interest on a loan or advance, for example, when the borrower is more than a particular period in arrears with respect to the payment of interest or principal. A bank discloses the aggregate amount of loans and advances at the balance sheet date on which interest is not being accrued and the basis used to determine the carrying amount of such loans and advances. It is also desirable that a bank discloses whether it recognizes interest income on loans and advances, and the impact which the non-accrual of interest has on its income statement.

When loans and advances cannot be recovered, they are written off and charged against the provision for losses. In some cases, they are not written off until all the necessary legal procedures have been completed and the amount of the loss is finally determined. In other cases, they are written off earlier, for example when the borrower has not paid any interest or repaid any principal that was due in a specified period. As the time at which uncollectable loans and advances are written off differs, the gross amount of loans and advances and of the provision for losses may vary considerably in similar circumstances. As a result, a bank discloses its policy for writing off uncollectable loans and advances.

Suggested Outlines

The following outlines suggested for the balance sheet, income statement and cash flow statement, are not intended to be prescriptive or indeed comprehensive. Since the circumstances and needs of banks will vary, these outlines are intended to provide a broad guide to what such statements contain. Banks use differing methods for the recognition and measurement of items in their financial statements. As a minimum, reference must be made to accounting policies on the following matters:

(a) recognition and treatment of interest income, particularly in relation to loans which are in default or doubtful;

(b) the basis of valuing foreign exchange assets and liabilities and the basis of inclusion of foreign exchange trading results in the income statement;

(c) valuation of all securities and other instruments, in particular the differentiation between those held for dealing and those for the longer term (investment);

(d) the manner in which provisions for bad and doubtful loans are determined.

Suggested Outline for the Balance Sheets of an Agricultural Bank
Balance Sheets as of 31 December (in $ million)

	1988	1989	1990
Assets			
Cash and balances with banks	78	80	75
Money market	70	70	70
Government securities	69	69	69
Other securities	-	-	-
Investments	-	-	-
Loans & bills discounted (gross)	549	641	698
Less provision for possible loan losses	94	115	137
Net loans and bills	455	526	561
Equity investments (in subsidiaries and others)	-	-	-
Less provision for possible losses	-	-	-
Net equity investments	-	-	-
Fixed assets (net)	12	11	20
Other assets	-	-	3
Total Assets	**684**	**756**	**798**
Liabilities & Shareholders' Equity			
Liabilities			
Deposits	67	85	121
Due to banks: Central Bank	250	300	300
Others	-	-	-
Bills payable	10	10	10
Provision for taxation	-	-	-
Other liabilities	1	2	2
Sub-Total	**328**	**397**	**433**
Shareholders' Equity			
Share capital	150	150	150
Reserves	22	22	22
Retained earnings	14	17	16
Government grants	170	170	177
Total Liabilities & Shareholders' Equity	**684**	**756**	**798**
Capital Adequacy Ratios	**(%)**	**(%)**	**(%)**
Equity 1/ ÷ total assets	52	47	46
Equity 1 ÷ risk assets	78	68	65

1/ Including Government grants.

Suggested Outline of Income Statements of an Agricultural Bank for Years ended 31 December (in $ million)

	1988	1989	1990
Interest income			
Interest on loans	40	49	54
Income from investment	-	-	-
Interest expenses (on deposits and borrowings)	11	14	16
Net interest income	**29**	**35**	**38**
Provision for possible loans losses	20	21	22
Net interest income after provision	9	14	16
Other operating income			
- Commission and fees	5	5	5
- Dividends from trade investments	-	-	-
Operating expenses			
- Staff	6	8	10
- Other expenses 1/	7	8	12
Profit before taxation	1	3	(1)
Taxation 2/	-	-	-
Profit after taxation	-	-	-
Dividends	-	-	-
Retained earnings	1	3	(1)
Ratios	**(%)**	**(%)**	**(%)**
(a) As % of average total assets:			
(i) Interest income	5.8	6.8	6.9
(ii) Net interest income	4.2	4.9	4.9
(iii) Net interest income after provisions	1.3	1.9	2.1
(iv) Operating expenses	1.9	2.2	2.8
(v) Net profit after tax	0.1	0.4	(0.1)
(b) Interest expenses as % of average total liabilities	3.4	3.9	3.9
(c) Interest income as % of average total loans outstanding	7.3	8.2	8.1
(d) Interest margin (c - b)	3.9	4.3	4.2
(e) Net profit as % of shareholders' equity at the beginning of the year	0.5	1.6	(0.5)

1/ Includes depreciation of $2 M, $2 M and $3 M for 1988, 1989 and 1990 respectively.
2/ Exempted from tax.

Suggested Outline for the Cash Flow Statement for an Agricultural Bank
Project Cash Flows for the Year ended 31 December (in $ million)

	1988	1989	1990
Cash Inflow			
Loan collection: - Principal - Interest	 127 23	 178 40	 229 47
Other income	5	5	5
Borrowings	88	76	51
Government Grants	-	-	7
Total Cash Inflow	**243**	**299**	**339**
Cash Outflow			
Loan disbursement	253	261	278
Repayment borrowings: - Principal - Interest	 4 11	 8 14	 15 16
Purchase of assets	-	-	16
Operating expenses	11	14	19
Total Cash Outflow	**279**	**297**	**344**
Surplus/(Deficit)	(36)	2	(5)
Opening cash balance	114	78	80
Closing cash balance	78	80	75

An Agricultural Bank
Capital Adequacy Analysis as at 31 December 1990 (in $ million)
(Current Assessment of Categorized Assets)

Assets at Risk as at 31 December 1990	**Total Amount of Portfolio**	**Without Risk Assets**	**Low Risk Assets**	**Moderate Risk Assets**	**High Risk Assets**	
Cash & dues from Banks						
Money market paper & tradeable securities						
Government securities						
Performing assets						
Non-performing assets						
Sub-Total						
Average Risk Factors						
A. Average Capital Requirement - Maximum Risk Factors						
B. Maximum Capital Requirement						
Resources Available for Coverage	**Total**	**Paid-in Shares**	**Reserves**	**Government Grants**	**Provision for Losses**	**Fixed Assets, Furniture Equipment**
Equity, Reserves, Provisions Assets at Cost: Fixed Leased C. Net Availability						

Coverage - Average (C/A) .... (times); Coverage - Maximum (C/B)..................(times).

Suggested Format for Calculation of Debt Service and Interest Cover Ratios for a Development Finance Institution

Debt Service and Interest Cover Ratios

	1988	1989	1990	......	......	......
Cash Receipts						
1. Net profit after tax						
2. Add back:						
(a) Non-cash charges (depreciation, provision for losses)						
(b) Accrued interest income at the start of the year 1/						
(c) Sub-total (a + b)						
(d) Deduct accrued interest income at the end of the year 1/						
(e) Sub-total (c - d)						
3. Cash loan collections (principal only)						
4. Interest expenses						
5. Total Cash Receipts (1 + 2e + 3 + 4)						
Debt Servicing						
6. Interest expenses (same as 4. above)						
7. Repayment of loans and borrowings						
8. Total debt servicing (6 + 7)						
9. Debt service cover (5 ÷ 8)						

1/ This adjustment is made to estimate income actually collected.

APPENDIX 1

FINANCIAL FORECAST

The effects of investment decisions are analysed through forecast of all three statements - Balance Sheet, Profit and Loss Account, and Source and Application of Funds Statement. Various methods that could be used to prepare projections are first discussed, followed by a description of individual items to be forecast in the individual statements.

Sales Forecasts

Sales potential generally becomes the limiting factor of company expansion. A good sales forecast therefore is an essential initial requirement. Given the sales forecast, financial projections can be attempted using methods such as **percent of sales, scatter diagram** or **regression.**

Percent of sales method. This method assumes that certain balance sheet items vary directly with sales. This means that the ratio of a given balance sheet item to sales remains constant and has a linear relationship. This method will be useful only for short-term forecasting since the relationship between sales and other variables is likely to change over time and there is no certainty that the linear relationship would hold in the longer term.

As an example, consider a cotton seed company whose balance sheet as at 31.12.89 is shown below. The company's sales are averaging $200,000 a year at full capacity. Profit after tax as a percentage of sales is 5%. During 1989, the company earned $10,000 after taxes and distributed $5,000 as dividends. How much additional funds will it require if sales expand to $300,000 in 1990.

As implied in the balance sheet, a higher level of sales causes more cash, more trade debtors, greater inventory and additional plant capacity among assets; and higher trade credits and larger accrued expenses among liabilities. The change in retained earnings will, however, depend on the amount of profits made and the portion distributed. Similarly, share capital and long-term loans may not move in the same direction as sales.

Balance Sheet Items
(expressed as % of sales - 31.12.89)

Capital, Reserves and Liabilities	**%**	**Assets**	**%**
- Share capital	n.a.	Net fixed assets	40
- Retained earnings	n.a.	Inventory	15
- Long-term loan	n.a.	Trade debtors	16
- Trade creditors	11	Cash	4
- Accrued expenses	4		
Total	**15**	**Total**	**75**

n.a. = not applicable

Assets as a % of sales	= 75%
Less relevant liabilities	= 15%
Percentage of each additional dollar of sales to be financed	= **60%**

The above calculations indicate that an additional $60,000 (60% x $100,000) will have to be financed, partly from profits ($7,500) and partly ($52,500) from equity capital or borrowing or a combination of both. Total sales in 1990 are $300,000, which yield a profit after tax of $15,000 of which $7,500 will be distributed as dividends leaving retained earnings of $7,500.

Scatter graph. Alternative methods used for forecasting financial requirements include the scatter graph or simple regression analysis. This is a better method particularly for longer-term forecasts, relating sales to other variables. A more sophisticated approach to forecasting a firm's assets involves the use of **multiple regression analysis,** in which sales are assumed to be a function of a number of variables unlike in simple regression where sales are related to only one variable. These methods are outside the scope of this paper and the interested reader may refer to standard books on economic or business statistics.

There are other more elaborate methods for projecting financial statements. Statements can be prepared based on functional budgets, or individual asset/liability estimation. Procedures involved are described below.

Balance Sheet Projections

The assets and the liabilities of a firm are usually estimated on the bases indicated below.

Trade debtors. Trade debtors at a particular date can be estimated by adding to the trade debtors at beginning of the period the total projected credit sales during the intervening period less total projected credit collection for the period. The latter can be ascertained from the cash budget.

If a cash budget is not available, the trade debtor balance may be estimated on the basis of a **turnover ratio**. This ratio, showing the relationship between credit sales and trade debtors, should be based on past experience. To obtain the estimated level of trade debtors, projected sales are divided by the turnover ratio.

Stocks. The estimated investment in stocks at a particular date may be based upon the production schedule which in turn is based on forecast sales. This schedule should show expected purchases, the expected use of stocks in production and the expected level of finished goods. On the basis of this information together with the beginning stock level, an estimate of stock can be made. Rather than use the production schedule, estimates of future stock can be based on a turnover ratio of cost of goods sold to stock. This ratio is applied in the same manner as for trade debtors.

Net fixed assets. These are estimated by adding planned expenditures to existing net fixed assets and deducting from this the sum of depreciation for the period plus any sale of fixed assets at book value.

Trade creditors. These are estimated by adding to opening balance on trade creditors the total projected purchases for the period less total projected cash payments for purchases during the period. If a cash budget is not available, a similar approach to that adopted to derive trade debtors can be used. To obtain the estimated level of trade creditors, projected purchases are divided by the (creditors) turnover ratio.

Accrued wages and expenses. The calculation is based on the production schedule and the historical relationship between these accruals and production.

Taxes payable. This is estimated by adding to the current balance, taxes on forecast income for the period less the actual payment of taxes.

Share capital and reserves. Share capital and reserves at the end of period would be the relative figure at beginning plus profits after taxes for the period less the cash dividends paid. If there are new issues of shares, then these would be added at issue price to the share capital to derive the final balance.

Cash and bank balances. A cash budget will show the estimated cash and bank balances. In general, cash serves as the balancing factor in the preparation of projected balance sheets.

If a cash budget is not available, one can prepare a projection of the balance sheet by making direct estimates of all the items by projecting financial ratios into the future and then making estimates on the basis of these ratios. Trade debtors, stocks, trade creditors, and accrued wages and expenses frequently are based on historical relationships to sales and production.

Projected Profit and Loss Statement

Sales and cost of goods sold. As is true with other projections, the sales forecast is the key input for projecting profit and loss statements. Given this forecast, production schedules can be formulated and estimates made of production costs for the product or products. An analyst can evaluate each component of the cost of goods sold - purchases, production, wages and overhead costs. Often however, costs of goods are estimated on the basis of ratios of the cost of goods sold to value of sales.

Administrative and marketing expenses. Because both of these expenses usually are budgeted in advance, estimates of them can be fairly accurate. Typically, these expenses are not overly sensitive to changes in sales in the very short-term, particularly to reduction in sales.

Company taxes. Taxes are computed based on applicable tax rates. Like the balance sheet, the projected profit and loss statement can be prepared on the basis of direct estimates rather than through the cash budget.

Cash Budget

A cash budget involves a projection of future cash receipts and cash disbursements of the firm over various intervals of time (normally within a year). It reveals the timing and the amount of expected cash inflows and outflows over the period studied. This information would help to determine the future cash needs of the firm, to plan for the financing of these needs and exercise control over the cash and liquidity of the firm in the short-term.

The forecast of sales. The key to the accuracy of most cash budgets is again the forecast of sales. This forecast is based on an internal analysis, an external one or both. The final sales forecast should be based on prospective demand and not be modified initially by internal constraints, such as physical capacity. The decision to remove these constraints will depend on the forecast.

Collections and other cash receipts. Given the sales forecast, the next objective is to determine the cash receipts from these sales. With cash sales, cash is received at the time of sale. With credit sales, however, the receipts do not come until later. The rate of inflow would depend on the terms of credit, the type of customer and the credit and collection policies of the firm.

Cash receipts may also arise from the sale of assets. For the most part, these are planned in advance and therefore are easily predicted for purposes of cash budgeting.

Cash disbursements. Given the sales forecast, a production schedule can be established. Firms may choose either to gear production closely to sales or to produce at a relatively constant rate over time. With the former production strategy, inventory carrying costs generally are lower but total production costs are higher than with the latter strategy. With steady production, the opposite usually occurs. Once a production schedule has been established, estimates can be made of the materials that will need to be purchased, the labour that will be required and the additional fixed assets the firm will need to acquire. As with trade debtors, there is a lag between the time a purchase is made and the time of actual cash payments.

Wages are assumed to increase with the amount of production. Generally wages are more stable over time than purchases. When production falls slightly, workers are usually not laid off. Included in other expenses are: general administrative and marketing expenses. These expenses tend to be reasonably predictable over the short run.

Other disbursements. These include capital expenditures, dividends, company taxes and any other cash outflows. Because capital expenditures are planned in advance, they are usually predictable for the short-term cash budget. Dividend payments are generally stable and are paid on specific dates. Estimation of company tax must be based on projected profits for the period under review.

PROFORMA STATEMENT

Cash Budget - Receipts and Payment Methods

Receipts

Cash sales
Receipts from debtors
Investment income
Sale proceeds of fixed assets

Payments

Matcrials and services
Wages and salaries
Capital expenditure
Interest payments
Loan instalment
Tax
Dividend

Surplus (or deficit)
Cash and bank balance at beginning
Cash and bank balance at end

Range of Cash Flow Estimates

In the face of uncertainty, we must provide information about the range of possible outcomes. To analyse cash flows only under one set of assumptions, as is the case with conventional cash budgeting, results in an incomplete perspective of the future.

GLOSSARY OF ACCOUNTING TERMS

General Terminology

The following glossary of accounting terms is not comprehensive but is intended to provide broad definition of some of those terms that are referred to in the preceding paper & some of those that financial analysts may come across in practice.

Absorption costing: A principle whereby fixed as well as variable costs are allotted to cost units and total overheads are absorbed according to activity level. The term may be applied where (a) production costs only or (b) costs of all functions are so allotted. See figure 1.

Accounting:

1. The classification and recording of actual transactions in monetary terms, and

2. The presentation and interpretation of the results of those transactions in order to assess performance over a period, and the financial position at a given date, and

3. The projection in monetary terms of future activities arising from alternative planned courses of action.

Functional relationships are illustrated in Figure 2.

Accounting bases: The methods developed for applying fundamental accounting concepts to financial transactions and items, for the purpose of financial accounts and in particular (a) for determining the accounting periods in which revenue and costs should be recognised in the profit and loss statement and (b) for determining the amounts at which material items should be stated in the balance sheet.

Accounting concepts: The basic assumptions which underlie the periodic financial accounts of business enterprises and may affect the way in which accounting information is collected, processed and disseminated. These include accrual/matching concept, consistency concept, entity concept, going concern concept, money measurement concept, prudence concept, realisation concept, etc.

Accounting policies: The specific accounting bases selected and consistently followed by a business enterprise as being, in the opinion of the management, appropriate to its circumstances and best suited to present fairly its results and financial position.

Accounting standards: General guidelines established by authority, custom or general consent which serves as a model for the practice of accounting.

Accrual accounting: A form of accounting wherein revenues and costs are matched, one with the other, and dealt with in the profit and loss statement of the period to which they relate irrespective of the period of receipt or payment.

Accrued expenses: Costs relating to a period which have not so far been taken into account because they have not yet been invoiced by the supplier or paid.

Administration cost: Cost of management and of secretarial, accounting and administrative services which cannot be directly related to the production, marketing, research or development functions of the enterprise.

Amortization: This term is used in two different contexts. The first usage relates to the retirement of debt on an installment or serial payment basis. The second usage relates to the provision for the using up of a wasting asset; e.g. oil fields, mines or quarries or of a fixed asset which has a predetermined useful life, e.g leases or intangible assets like goodwill, patents and trade marks.

Annual report and accounts: A set of statements comprising a management report (in the case of companies the director's reports), the balance sheet, profit and loss account or other revenue account and related notes.

Asset: Any possession of value.

Asset cover: Net tangible assets before deducting borrowings divided by total borrowings. This indicates the safety of the lender's money.

Asset value per share: Equity share capital plus reserves divided by the number of issued equity shares. This shows the value of assets per share, usually for the benefit of equity shareholders.

Associated company: A company not being a subsidiary of the investing group or company in which

a) the interest of the investing group or company is effectively that of a partner in a joint venture or consortium and the investing group or company is in a position to exercise a significant influence over the company in which the investment is made; or

b) the interest of the investing group or company is for the long term, and is substantial and, having regard to the disposition of the other shareholdings, the investing group or company is in a position to exercise a significant influence over the company in which the investment is made.

Attributable profit (on contracts): That part of the total profit currently estimated to arise over the duration of the contract (after allowing for likely increases in costs so far not recoverable under the terms of the contract) which fairly reflects the profit attributable to that part of the work performed at the accounting date. There can be no attributable profit until the outcome of the contract can be assessed with reasonable certainty.

Audit: A systematic examination of the activities and status of an entity based primarily on investigation and analysis of its systems, controls and records. External audit is a periodic examination of the books of account and records of an entity carried out by an independent third party (the auditor), to ensure that they have been properly maintained, are accurate and comply with established concepts, principles, accounting standards, legal requirements and give a true and fair view of the financial state of the entity. Internal audit is an independent appraisal activity established within an organization as a service to it. It is a control which functions by examining and evaluating the adequacy and effectiveness of other controls. Management audit is an objective and independent appriasal of the effectiveness of managers and the effectiveness of the corporate structure in the achievement of company objectives and policies. Its aim is to identify existing and potential management weaknesses within an organization and recommend ways to rectify these weaknesses.

Audit report:

1. A report by an auditor in accordance with his terms of appointment.

2. The formal document in which an auditor expresses his opinion as to whether, in accordance with accepted accountancy principles and legal requirements, the financial statements of an entity show a true and fair view of:

   a) its position at a given date and

   b) the results of its operations for the accounting period ended on that date.

Auditing standards: Those statements of auditing standards which are approved for issue by the regulatory bodies and which are effective for the period covered by the financial statements on which the auditor is reporting.

Back to back loan: A form of financing whereby money borrowed in one country or currency is covered by the lending of an equivalent amount in another.

Bad debt: A debt which is or is considered to be uncollectable and is, therefore, written off as a charge to the profit and loss account. Bad debt ratio is bad debts incurred divided by sales on credit or bad debts incurred divided by total debtors at a point in time. Both ratios highlight the effectiveness of credit control.

Bad debt provision: A financial provision representing an estimate of uncollectables of accounts receivable.

Balance sheet: A statement of the financial position of an entity at a given date disclosing the value of assets, liabilities and accumulated funds such as shareholders' contributions and reserves, prepared to give a true and fair view of the state of the entity at that date.

Bank reconciliation: A detailed statement reconciling, at a given date, the cash balance in an entity's cash book with that reported by a bank in a bank statement.

Bankruptcy: The legal status of an individual against whom an adjudication order has been made by the court primarily because of his inability to meet his financial liabilities.

Batch costing: That form of specific order costing which applies where similar articles are manufactured in batches either for sale or for use within the undertaking.

Bill payable: A bill of exchange or promissory note payable, prepared by a drawer to a specified payee and often subject to acceptance by a third party as a guarantor or as a discounter.

Bill receivable: A bill of exchange or promissory note receivable.

Bonus or scrip issue: The capitalisation of the reserves of a company by the issue of bonus shares to existing shareholders, in proportion to their holdings, such shares being normally fully paid-up with no monetary payments being made.

Book-Keeping: That part of accounting which deals with the recording of actual transactions in monetary terms. This may be in books, or on cards, tape, disc or a combination of these.

Book value: The historical cost of an asset. Net book value is the historical cost of an asset less any accumulated depreciation or other provision for diminution in value.

Branch accounts: A term used for a form of departmental accounts where different geographical locations are involved; usually applied in retail businesses and also to branches in other countries where branches are not incorporated.

Break-even analysis: An analytical approach for illustrating the relationship between fixed cost, variable cost, sales and profit. The analysis may employ a break-even chart which indicates approximate profit or loss at different levels of sales volume. The break-even point can be ascertained by various methods including the use of a break-even chart. The break-even point may also be calculated by formulae, as follows:

$$\frac{\text{Total Fixed Cost}}{\text{Contribution per Unit}} = \text{Number of units to be sold to break even}$$

$$\frac{\text{Total Fixed Cost x Sales Value}}{\text{Total Contribution}} = \text{Sales value at break even point}$$

Budget: A plan quantified in monetary terms, prepared and approved prior to a defined period of time, usually showing planned income to be generated and/or expenditure to be incurred during that period and the capital to be employed to attain a given objective.

Budget centre: A section of an organization for which separate budgets can be prepared and control exercised.

Budgetary control: The establishment of budgets relating the responsibilities of executives to the requirements of a policy, and the continuous comparison of actual with budgeted results, either to secure by individual action the objective of that policy or to provide a basis for its revision.

Capital commitment:

a) The aggregate amount or estimated amount of contracts for capital expenditure, so far as not provided for; and

b) the aggregate amount or estimated amount of capital expenditure authorised by the directors which has not been contracted for.

Capital employed: The funds used by an entity for its operations. This can be expressed at various levels according to purpose of use.

		$'000	
Fixed assets		1,000	
Investments		200	
Working capital			
Stock	100		
Trade debtors and prepayments	80		
Cash & bank balances	20		
	200		
Less trade creditors & accruals	(50)	150	
Operations management capital employed		1,350	A
Less tax creditor		(25)	
Less dividends payable		(25)	
Company capital employed		1,300	B
Less borrowings		(600)	
Plus goodwill		40	
Shareholders' capital employed		740	C
		===	
Financed by:			
Share capital		500	
Reserves		240	
		740	
		===	

A and B to be related to pre-tax and pre-interest profits. C to be related to profits after tax.

Capital expenditure: Expenditure on fixed assets or additions thereto intended to benefit future accounting periods or expenditure which increases the capacity, efficiency, life span or economy of operation of an existing fixed asset.

Capital funding planning: The process of selecting suitable funds to finance long-term assets and working capital.

Capital gain (loss): The extent by which the net realised value of a capital asset exceeds (or in the case of a capital loss is less than) the cost of acquisition plus additional improvements less depreciation charges where applicable.

Capital structure: This refers to the make-up of the long-term capital sources represented by long-term debt, share capital and reserves.

Capital surplus: The surplus distributed amongst shareholders in accordance with their rights under the articles of association after the discharge of all outstanding costs and liabilities following the liquidation of a company.

Capital turnover: Turnover of the year divided by the average capital employed in the year. This measures the number of times the capital is turned over in the year. Also known as assets turnover.

Cash accounting: A form of accounting wherein transactions are recorded only as cash is paid or recieved and related financial statements are usually restricted to a summary of receipts and payments.

<u>Cash flow</u>: The cash generated or spent in a given period.

<u>Cash flow statement</u>: A statement produced for the management of an entity showing, by broad category, the cash received and spent for a given period.

<u>Concept - accrual/matching</u>: The concept that revenues and costs are matched one with the other and dealt with in the profit and loss account of the period to which they relate irrespective of the period of receipt or payment. See also accrual accounting.

<u>Concept - business entity</u>: The concept that financial accounting information relates only to the activities of the business entity and not to the activities of its owner(s).

<u>Concept - consistency</u>: The principle that there is a consistency of treatment of like items (accounting bases and policies) within each accounting period and from one period to the next.

<u>Concept - going concern</u>: The assumption that the enterprise will continue in operational existence for the foreseeable future.

<u>Concept - lower of cost or net realisable value</u>: a concept of stock valuation whereby goods in stock are valued at actual or standard cost or at replacement cost or net realisable value whichever is lower.

<u>Concept - materiality</u>: The principle that financial statements should separately disclose items which are significant enough to affect evaluation or decisions. The level of significance is a matter for individual judgement.

<u>Concept - money measurement</u>: The concept that financial accounting information relates only to those activities which can be expressed in monetary terms.

<u>Concept - prudence</u>: The concept that revenue and profits are not anticipated, but are recognised by inclusion in the profit and loss account only when realised in the form either of cash or of other assets, the ultimate cash realisation of which can be assessed with reasonable certainty; provision is made for all known liabilities (expenses and losses) whether the amount of these is known with certainty or is a best estimate in the light of the information available.

<u>Concept - realisation</u>: The concept that profit is only accounted for when it is realised and not when it can be recognised.

<u>Consolidated financial statements/group accounts</u>: A form of group accounting which presents the information contained in the separate financial statements of a holding company and its subsidiaries as if they were the accounts of a single entity.

<u>Contingent liabilities</u>: Liabilities which are dependent on a condition which exists at the balance sheet date, where the outcome will be confirmed only on the occurrence or non-occurrence of one or more uncertain future events.

<u>Contra</u>: A book-keeping term meaning against, or on the opposite side. It is usually used where debts are matched with related credits, in the same or a different account.

Contract costing: That form of specific order costing which applies where work is undertaken to customers' special requirements and each order is of long duration.

Contribution: The difference between sales value and variable cost of those sales expressed either in absolute terms or as a contribution per unit.

Contribution centre: A profit centre where expenditure is calculated on a marginal cost basis.

Contribution to sales ratio: Contribution divided by sales and expressed as a percentage; also known as profit/volume ratio.

Conversion cost: Cost of converting material input into semi-finished or finished products, i.e. additional direct materials, direct wages, direct expenses and absorbed production overheads.

Convertible loan stock: A loan which gives the holder the right to convert to other securities, normally ordinary shares, at a pre-determined rate and time.

Cost: The amount of expenditure (actual or notional) incurred on or attributable to, a specified thing or activity. The word can rarely stand alone and should be qualified as to its nature and limitations.

Cost accounting: That part of management accounting which establishes budgets and standard costs and actual costs of operations, processes, departments or products and the analysis of variances, profitability or social use of funds.

Cost allocation: The charging of discrete identifiable items of cost to cost centres or cost units.

Cost apportionment: The division of costs amongst two or more cost centres in proportion to the estimated benefit received, using a proxy, e.g. rent in proportion to area.

Cost ascertainment: The collection of costs attributable to cost centres and cost units using the costing methods, principles and techniques prescribed for a particular business entity.

Cost audit: The verification of cost records and accounts and a check on the adherence to the prescribed cost accounting procedures and the continuing relevance of such procedures.

Cost centre: A location, function or items of equipment in respect of which costs may be ascertained and related to cost units for control purposes.

Cost of capital: The cost of financing an investment, expressed as a percentage rate. The rate should be based on the overall pool of capital. Weighted average cost of capital is the average cost of the combined sources of finance (equity, loans) weighted according to the proportion each element bears to the total pool of capital available. Weighting is usually based on the current market valuation and current yields or costs. Example:

Capital	Market value	Rate	Cost
Equity	$1,000	8%	$80
Loan	$1,000	10%	$100

$$\text{Weighted average} = \frac{180}{2,000} \times 100 = 9\%$$

Cost of sales: The sum of direct cost of sales plus factory overheads attributable to the turnover.

Cost unit: A quantitative unit of product or service in relation to which costs are ascertained.

Creditors (accounts payable): Money owed to suppliers or others.

Creditor days ratio: Average trade creditors or at end of period divided by average daily purchases on credit terms. This ratio measures the average time taken, in days, to pay for supplies received on credit.

Credit report: A report giving information about an individual or company which may bear on a decision to grant credit to that individual or company.

Current asset: Cash or other asset, e.g. stock, debtors or short-term investment, held for conversion into cash in the normal course of trading.

Current cost: The calculated cost of acquiring goods for processing or resale, or of assets, at the time when their value is consumed by the entity, usually obtained by some form of averaging or index.

Current cost accounting: A system of accounting based on a concept of capital which is represented by the net operating assets of a business. These net operating assets (fixed assets, stocks and monetary working capital) are the same as those included in historical cost accounting but in the current cost accounts the fixed assets and stock are normally expressed at current price levels.

Current expenditure: Also known as revenue expenditure or recurrent expenditure. Expenditure on the supply and manufacture of goods and provision of services charged in the accounting period in which they are consumed.

Current liabilities: Liabilities which fall due for payment in a relatively short period normally less than twelve months, e.g. creditors, current taxation, dividends payable; also, that part of long-term loans due for repayment within one year.

Current purchasing power accounting: A system of accounting for inflation in which the values of the non-monetary items in the historical cost accounting statements are adjusted to reflect changes in the general purchasing power of money. Adjustments are made using a general price index.

Current ratio: Current assets at end of period divided by current liabilities at end of period. This ratio is a measure of an overall test of liquidity.

Debtors' age analysis: An analysis of sums owing by debtors, classified according to age of debt.

Debt capacity: The extent to which an entity can raise loan finance.

Debtors' days ratio: Average trade debtors or at end of period divided by average daily sales (turnover) on credit terms in the period. This ratio measures the debtors outstanding at the end of the period in terms of average daily sales on credit in the period. Example:

$$\frac{\text{Trade debtors}}{\text{Average daily sales on credit}} \quad \frac{\$10,000}{500} = 20 \text{ days}$$

Debenture: The written acknowledgement of a debt by a company, usually given under seal, and normally containing provisions as to payment of interest and the terms of repayment of principal. A debenture may be secured on some or all of the assets of the company. An unsecured debenture is known as a naked debenture.

Debt/equity ratio: Also known as gearing/financial leverage ratio. The relationship between shareholders' capital plus reserves, and either prior charge capital or borrowings or both, as follows:

1. $\dfrac{\text{Prior charge capital (A)}}{\text{Total capital in issue plus reserves (B)}}$

2. $\dfrac{\text{Total borrowings (C)}}{\text{Total capital in issue plus reserves}}$ or $\dfrac{\text{Total borrowings}}{\text{Ordinary share capital in issue plus reserves}}$

3. $\dfrac{\text{Prior charge capital plus borrowings}}{\text{Total capital in issue plus reserves}}$

4. $\dfrac{\text{Prior charge capital}}{\text{Capital employed (D)}} \times 100$

'A' includes preference shares and debentures.
'B' includes prior charge capital plus equity capital and reserves.
'C' combines total long-term and short-term borrowings.

Ratio 4 above can also be expressed by the formula: $\dfrac{\text{Prior charge capital}}{\text{Equity}}$

Debt factoring: The sale of debts to a third party (the factor) at a discount, in return for prompt cash.

Debt service: The aggregate amount of amortization (including sinking fund payments if any) of, and interest and other charges on, debt.

Deferred expenditures (charges): Expenditure not charged against income in an accounting period but carried forward to be charged in the next or a subsequent period, e.g. advertising expenditure where the benefit is expected to be received in a future period.

Deficiency (or surplus) account: A statement in a prescribed form, which shows the excess, if any, of liabilities over assets (or vice versa) at a given date.

<u>Depreciation</u>: The measure of the wearing out, consumption or other loss of value of a fixed asset whether arising from use, effluxion of time or obsolescence through technology and market changes.

<u>Departmental accounts</u>: Accounts which show the revenue and expenditure of the various departments of an entity for a given period. They may take the form of a manufacturing, trading and profit and loss account for each department but could also be an operating account for a service department.

<u>Development cost</u>: The cost of using scientific or technical knowledge in order to produce new or substantially improved materials, devices, products, processes, systems or services prior to the commencement of commercial production.

<u>Differential costing</u>: A technique used in the preparation of ad hoc information in which only cost and income differences between alternative courses of action are taken into consideration.

<u>Direct cost of sales</u>: The sum of direct materials consumed, direct wages, direct expenses and variable production overheads.

<u>Direct expenses</u>: Costs other than materials or labour, which can be identified in a specific product or saleable service.

<u>Direct labour cost</u>: The cost of remuneration for employees' efforts and skills applied directly to a product or saleable service and which can be identified separately in product costs.

<u>Direct materials cost</u>: The cost of materials entering into and becoming constituent elements of a product or saleable service and which can be identified separately in product cost.

<u>Distribution cost</u>: Cost incurred in warehousing saleable products and in delivering products to customers.

<u>Dividend</u>: A distribution to shareholders out of profits, usually in the form of cash.

<u>Dividend cover</u>: Earnings per share divided by net dividend per share. This indicates the number of times the profit available to the equity shareholders covers the actual net dividends payable for the period.

<u>Earnings per share</u>: Attributable equity profit for the period divided by the number of equity shares in issue and ranking for dividends.

<u>Entity</u>: An economic unit that has a separate, distinct identity.

<u>Equity</u>: Usually the issued ordinary share capital plus attributable reserves.

<u>Equity share capital</u>: Normally the ordinary shares of a company.

<u>Exceptional items</u>: Items of abnormal size and incidence which are derived from the ordinary activities of the business.

<u>Extraordinary income and charges</u>: Income or cost which derives from events or transactions outside the ordinary activities of the business and which are both material and expected not to occur frequently or regularly. They do not include items which, though exceptional on account of size and

incidence, derive from the ordinary activities of the business. Neither do they include prior year items merely because they relate to a prior year.

Fictitious asset: An item shown as an asset in the balance sheet which has no realisable value, e.g. preliminary and formation expenses of a company or special advertising.

Financial accounting: That part of accounting which covers the classification and recording of actual transactions of an entity in monetary terms in accordance with established concepts, principles, accounting standards and legal requirements and presents as accurate a view as possible of the effect of those transactions over a period of time and at the end of that time.

Financical control: The use of management accounting information, the comparison of planned and actual performance and taking action to correct adverse trends or optimise favourable conditions. This is not to be confused with control of finance, which is the main function of treasurership.

Financial leverage: The extent to which the company uses prior charge capital to finance its assets. See also debt/equity ratio.

Financial management: The process of financial decision-making based on the planning, forecasting, organising, controlling and communicating of financial and physical data derived from the design and implementation of a project, with the objective of achieving optimum financial and economic benefits from an investment.

Financial planning: Planning in monetary terms of the acquisition and financing of resources and their utilisation.

Financial statements are balance sheets, profit and loss accounts, statements of source and application of funds, notes and other statements, which collectively are intended to give a true and fair view of financial position and profit or loss.

Financial structure: Broader concept than capital structure, showing how firm's total assets are financed.

First in First out (FIFO): A method of pricing material issues using the oldest purchase price first.

Fixed asset: Any asset acquired for retention in an entity for the purpose of providing goods or services and not held for resale in the normal course of trading.

Fixed assets turnover ratio: Turnover of the year divided by the average net book value of fixed assets. This ratio measures the turnover generated by each $1 of fixed assets or the number of times fixed assets are turned over in the year.

Fixed budget: A budget which is designed to remain unchanged irrespective of the volume of output or turnover attained.

Fixed charge: A claim on a specific asset of an entity, given as debt security.

Fixed overhead cost: The cost which accrues in relation to the passage of time and which, within certain output and turnover limits, tends to be unaffected by fluctuations in the levels of activity. Other terms used include period cost and policy cost.

Flexible budget: A budget which, by recognising the difference in behaviour between fixed and variable costs in relation to fluctuations in output, turnover, or other variable factors such as the number of employees, is designed to change appropriately with such fluctuations.

Floating charge: A general claim on the assets of an entity, given as debt security, without attachment to a specific asset.

Funds flow: Broader concept than cash flow, focussing upon changes in working capital.

Fungible assets: Assets which are substantially indistinguishable one from another.

Gearing: The relationship between shareholders' capital plus reserves and either prior charge capital or borrowings or both.

Goodwill: The intangible benefit arising from the commercical connections and reputation of a business when a business is purchased; any amount paid in excess of its net worth represents the value placed on the goodwill.

Gross dividend yield: Actual dividend paid per share plus imputed tax divided by market price per share expressed in percentage. This ratio measures the percentage of current share price in terms of gross dividends.

Gross profit/margin: Turnover less cost of sales. Gross profit percentage is computed thus - $\frac{\text{Gross profit of the period} \times 100}{\text{Turnover of the period}}$

This ratio is commonly used to determine whether selling prices are adequate and to determine selling price policy.

Gross revenues: The total amount of revenues earned from all operational sources of an entity through the sales of products and/or provision of services during an accounting period.

Group: A holding company and its subsidiaries.

Historical cost: The actual cost of acquiring assets or goods and services.

Historical cost accounting: A system of accounting in which all values (in revenue and capital accounts) are based on the costs actually incurred or as revalued from time to time.

Holding gains: The difference between the measured value to a company of an asset at any point in time and the original cost incurred by the company in purchasing that asset (less depreciation where appropriate).

Income and expenditure account: An account (used by concerns such as clubs, associations, charities, whose main objective is not profit making) which shows the excess of income over expenditure (or vice versa) for a given period. The account is similar to a profit and loss account and is usually accompanied by a balance sheet.

Income statement: Also known as profit and loss account which shows the excess of revenues over expenses (or vice versa) for a given period.

Intangible assets: Any asset which does not have a physical identity, e.g. goodwill.

Indirect expenses: Expenses which are not charged directly to a product, e.g. insurance, water rates.

Indirect labour cost: Labour costs which are not charged directly to a product, e.g. supervision.

Indirect material cost: Material costs which are not charged directly to a product, e.g. cleaning materials.

Insolvency: The inability of a debtor to pay his debts as and when they fall due.

Interest cover: Profit before interest and tax divided by all interest payable. This ratio indicates the vulnerability of the interest payments to a drop in profits.

Inventories: Inventories comprise:

a) goods or other assets purchased for resale;

b) consumable stores;

c) raw materials and components purchased for incorporation into products for sale;

d) products and services in intermediate stages of completion;

e) finished goods.

Investment: Any application of money or money's worth which is intended to provide a return by way of interest, dividend or capital appreciation.

Investment centre: A profit centre in which inputs are measured in terms of expenses and outputs are measured in terms of revenues, and in which assets employed are also measured, the excess of revenue over expenses then being related to assets employed.

Job costing: That form of specific order costing which applies where work is undertaken to customers' special requirements and each order is of comparatively short duration.

Joint cost: The costs of providing two or more products or services whose production could not, for physical reasons, be segregated.

Last in First out (LIFO): A method of pricing material issues using the last purchase price first.

Liabilities: The financial obligations of a business, internal (e.g to shareholders) and external (e.g. to creditors).

Liquid assets: Cash and other assets readily convertible into cash.

Liquidation: The winding up of a company, in which assets are sold, liabilities settled as far as possible and any remaining cash is returned to the owners.

Liquidity ratios: A group of measures relating to working capital which indicate the ability to meet liabilities with the assets available.

Loan capital: Debentures and other long-term loans to a business.

Management accounting: The provision of information required by mnanagement for such purposes as:

1. formulation of policies;

2. planning and controlling the activities of the enterprise;

3. decision taking on alternative courses of action;

4. disclosure to those external to the entity (shareholders and others);

5. disclosure to employees;

6. safeguarding assets.

Manufacturing account: Establishes the production cost of the goods sold. The first part of the manufacturing account shows the prime cost while the second part shows production overheads. Production overheads are sometimes referred to as 'factory overheads', 'works overheads' or 'manufacturing overheads'.

Margin: An expression used to denote the difference in unit value percentage or total value, between realised sales and the cost of goods sold. Since the margin may be calculated at different stages, the terms gross and net margin are used to differentiate between the levels.

Marginal cost: The variable cost of one unit of a product or a service, i.e. a cost which would be avoided if the unit was not produced or provided.

Marginal costing: A principle whereby variable costs are charged to cost units and the fixed cost attributable to the relevant period is written off in full against the contribution for that period.

Marketing cost: The cost incurred in researching the potential markets and promoting products in suitably attractive forms and acceptable prices.

Merger: The amalgamation of two or more separate entities.

Minority interest: Shares held in a subsidiary company by members other than the holding company or its nominees plus the appropriate portion of the accumulated reserves.

Monetary working capital: The aggregate of

a) trade debtors prepayments and trade bills receivable, plus

b) stocks not subject to a cost of sales adjustment, less

c) trade creditors, accounts and trade bills payable

insofar as they arise from the day-to-day operating activities of the business as distinct from transactions of a capital nature.

Net assets: The excess of the book value of the assets of an entity over its liabilities including loan capital.

Net liquid funds: Cash at bank and in hand and cash equivalents, e.g. investments held as current assets, less bank overdrafts and other borrowings repayable within one year of the accounting date.

Net realisable value: The price at which goods in stock could be currently sold less any further costs which would be incurred to complete the sale.

Net worth: The paid-up share capital and reserves (less accumulated losses if any).

Off balance sheet finance: Sources of finance which do not appear on the balance sheet; generally the use of assets which are not owned, e.g. finance leasing.

Operating income/profit: Operating revenues less operating costs.

Operating leverage: The extent to which fixed costs are a component in a firm's cost structure.

Operating ratio: Gross revenues divided by operating expenses. This ratio shows the adequacy of revenues to meet the expenses.

Over/under capitalisation: Term used to describe a surplus or deficiency of permanent capital in relation to the current level of activity of a business.

Overhead cost: The total cost of indirect materials indirect labour and indirect expenses. Equivalent term in U.S.A. is burden.

Over-trading: A term applied to business which enters into commitments in excess of its available short-term liquid resources.

Parent company: A company that has one or more subsidiaries.

Payback: The period, usually expressed in years, which it takes the cash inflows from a capital investment project to equal the cash outflows.

Preference shares: Shares carrying a fixed rate of dividend, the holders of which, subject to the conditions of issue, have a prior claim to the repayment of capital in the event of winding-up.

Prepayments: Expenditure on goods and services for future benefit, which is to be charged to future operations, e.g. rent paid in advance.

Price/earnings ratio: Market price of share divided by earnings per share. This ratio shows the number of years it would take to recoup the investment in the share out of the earnings attributable. It is a reflection of the market's expectation of future earnings.

Prime cost: The total cost of direct materials, direct labour and direct expenses.

Prior year adjustments: Material adjustments applicable to prior years arising from changes in accounting policies and from the correction of fundamental errors. They do not include the normal recurring corrections and adjustments of accounting estimates made in prior years.

Prior charge capital: Those classes of share and loan capital, the holders of which have a claim on the profit and assets of a business before the ordinary shareholders.

Production cost: Prime cost plus production overheads.

Process costing: The basic costing method applicable where goods or services result from a sequence of continuous or repetitive operations or processes to which costs are charged before being averaged over the units produced during the period.

Product life cycle: The pattern of demand for a product or service over time.

Profit to turnover ratio: Profit before interest and tax divided by turnover expressed in percentage. This ratio uses turnover as the measure to determine profit performance.

Provisions for liabilities and charges: Amounts retained as reasonably necessary for the purpose of providing for any liability or loss which is either likely to be incurred, or certain to be incurred but uncertain as to amount or as to the date on which it will arise.

Qualified audit report: A statement in an audit report in which the auditor expresses reservations, doubts or exceptions regarding certain item(s) in the report or draws attention to a limitation in his examination, due to departures from generally accepted accounting principles or lack of consistency in their application or significant uncertainties affecting the financial statements or due to restrictions in the scope of the auditor's examination.

Quick assets: These comprise cash and those debtors, securities and other assets which can be quickly turned into cash.

Quick (asset) ratio: Also known as acid test ratio. Quick assets at the end of the period divided by current liabilities at the end of the period. This ratio measures the ability to pay creditors in the short term.

Rate of return (accounting): A ratio sometimes used in investment appraisal, which is analogous to the return on capital employed ratio. Unlike net present value and internal rate of return, the ratio is based on profits as opposed to cash flows. It is represented by the formula

$$\frac{\text{Average annual profit from investment x 100}}{\text{Average investment}}$$

Ratio pyramid: A part of ratio analysis whereby a primary ratio is broken down into secondary ratios which are mathematically linked. For example:

$$\text{Primary ratio} = \frac{\text{Profit}}{\text{Capital employed}}$$

$$\text{Secondary ratio} = \frac{\text{Profit}}{\text{Turnover}} \times \frac{\text{Turnover}}{\text{Capital employed}}$$

Receipts and payments accounts: An account which shows in summarised form the cash transactions during a given period.

Redeemable shares: Shares which are to be redeemed or are liable to be redeemed at the option of the company or the shareholder.

Redemption: Repayment, usually used in connection with repayment of preference shares or debentures.

Replacement cost: The cost at which an identical asset could be purchased or manufactured.

Replacement cost accounting: A family of techniques which seeks to show whether or not the value of physical assets of the business has been maintained in financial terms.

Research cost (applied): The cost of original investigation undertaken in order to gain new scientific or technical knowlege and directed towards a specific practical aim or objective.

Research cost (basic): The cost of original investigation undertaken in order to gain new scientific or tehcnical knowledge and understanding, not primarily directed towards any specific practical aim or application.

Reserves: Profits or surpluses which are retained in an entity. Some entities classify them into those which are distributable and those which are not distributable.

Responsibility accounting: A system of accounting that segregates revenues and costs into areas of personal responsibility in order to assess the performance attained by persons to whom authority has been assigned.

Responsibility centre: A unit or function of an organization headed by a manager having direct responsibility for its performance.

Retained earnings/profit: That part of profit for the financial year which is not distributed and is thus retained as revenue reserve/general reserve.

Return on capital employed: This measures the percentage profit generated as an indication of the productivity of capital employed. The ratio can be used for different measurement purposes according to constituent parts of average capital employed and definitions of profit.

Revaluation reserve (surplus): Excess value of assets over the latest value ascribed in the books of account often following a technical revaluation of the worth of the assets or as a result of indexing or other forms of adjustment to reflect changing prices or inflation.

Revenue centre: A centre devoted to raising revenue with no responsibility for production, e.g. a sales centre.

Revenue expenditure: See current expenditure.

Rights issue: The raising of new capital by a company, by giving existing shareholders the right to subscribe to new shares or debentures in proportion to their current holdings.

Rolling budget: The continuous updating of a short-term budget by adding, say, a further month or quarter and deducting the earliest month or quarter so that the budget can reflect current conditions.

Rolling forecast: A continuously updated forecast covering one or more periods ahead, whereby each time actual results are reported a further forecast period is added and intermediate period forecasts are updated.

Selling cost: Cost incurred in securing orders, usually including salesmens' salaries, commissions and travelling expenses.

Service cost centre: A centre devoted to provision of a service or services to other cost centres.

Service function costing: The costing of speciic services or functions, e.g. cafeteria, maintenance.

Share capital - authorised, nominal or registered: The type, class, number and amount of the shares which a company may issue.

Share capital - called up: The amount which the company has required shareholders to pay on the shares issued.

Share capital - issued or subscribed: The type, class, number and the amount of shares held by shareholders.

Share capital - paid up: The amount which shareholders are deemed to have paid on the shares issued and called up.

Share capital - uncalled: The amount of the share price of issued shares which has not been called up by the company.

Share capital - unissued: Ther amount of the share capital authorised but not issued.

Share premium: The excess paid to a company by a member either in cash or other consideration, over the nominal value of the shares issued.

Sinking fund: A fund created for the redemption of a liability, or with the object of replacing an asset, by setting aside a sum periodically, and investing it (usually outside the business) so as to produce the required amount at the appropriate time.

Source and application of funds statement: A statement showing the sources and values of funds flowing into an entity, the way in which they have been used and how any net surplus or deficiency in short and long-term funds has been applied.

Specific order costing: The basic costing method applicable where the work consists of separate contracts, jobs or batches, each of which is authorised by a specific order or contract.

Standard costing: A technique of cost accounting which compares the 'standard cost' of each product or service with the actual cost, to determine the efficiency of the operation, so that remedial action may be taken immediately.

Statement of affairs: A statement in a prescribed form, showing the estimated financial position of a debtor of a company which may be unable to meet its debts.

Stock turnover/number of days' stock held: Stock value divided by average daily cost of sales in period. This shows the number of days worth of stock held at the most recent rate of daily turnover. It can be applied to finished stocks, raw materials and work in progress by using appropriate numerators.

Number of weeks' stock: (finished goods) $\frac{\text{Finished goods stock}}{\text{Average weekly sales}}$

Number of weeks' stock: (raw materials) $\frac{\text{Raw material stock}}{\text{Raw material usage per week}}$

Number of weeks' stock: (work in progress) $\frac{\text{Work in progress}}{\text{Average weekly production}}$

Suspense account: An account in which debits or credits are held temporarily until sufficient information is available for them to be posted to the correct accounts.

Tangible assets: Any asset having a physical identity.

Trading account: An account which shows the gross profit/loss or contribution generated by an entity for a given period.

Transfer price: A price related to goods or other services transferred from one process or department to another or from one member of a group to another.

Treasurership: The function concerned with the provision and use of finance as distinct from the control function.

Trial balance: A list of debit and credit balances on the accounts, extracted at a given date.

Turnover: Amounts derived from the provision of goods and services falling within the company's ordinary activities, after deduction of trade discounts, value added tax and any other taxes on the amounts so derived.

Uniform accounting: A common system using agreed concepts, principles and standard accounting practice adopted by different entities in the same industry to ensure that they all deal with accounting information in a like manner, the objective being to facilitate inter-firm comparison.

Value added: The increase in realisable value resulting from an alteration in form, location or availability of a product or service excluding the cost of purchased material or services.

Variable cost: A cost which tends to follow the level of activity.

<u>Variable overhead cost</u>: Overhead cost which tends to vary with changes in the level of activity.

<u>Variance</u>: The difference between a budgeted or standard amount and the actual amount during a given period.

<u>Wasting asset</u>: Any asset of a fixed nature which is gradually consumed or exhausted in the process of earning income, e.g. mines, quarries.

<u>Weighted average cost of capital</u>: See cost of capital.

<u>Working capital</u>: The capital available for conducting the day-to-day operations of an organization; normally the excess of current assets over current liabilities.

<u>Work in progress</u>: Any material, component, product or contract at an intermediate stage of completion.

<u>Working ratio</u>: Most frequently used in transportation projects. It is the relationship between gross operating revenues from all operational sources to total operating expenses.

<u>Zero base budgeting</u>: A method of budgeting whereby all activities are re-evaluated each time a budget is formulated. Each functional budget starts with the assumption that the function does not exist and is at zero cost.

Figure 1

Costing Principles, Techniques and Methods
Costing principles
Costing Techniques
Absorption Costing
Marginal Costing
Standard Costing
Budgetary Control
Information Data Base
Costing Methods
Costing Methods
Specific order costing
Continuous operation/ process costing
Service/function costing
Contract costing
Job costing
Batch costing
Product costing

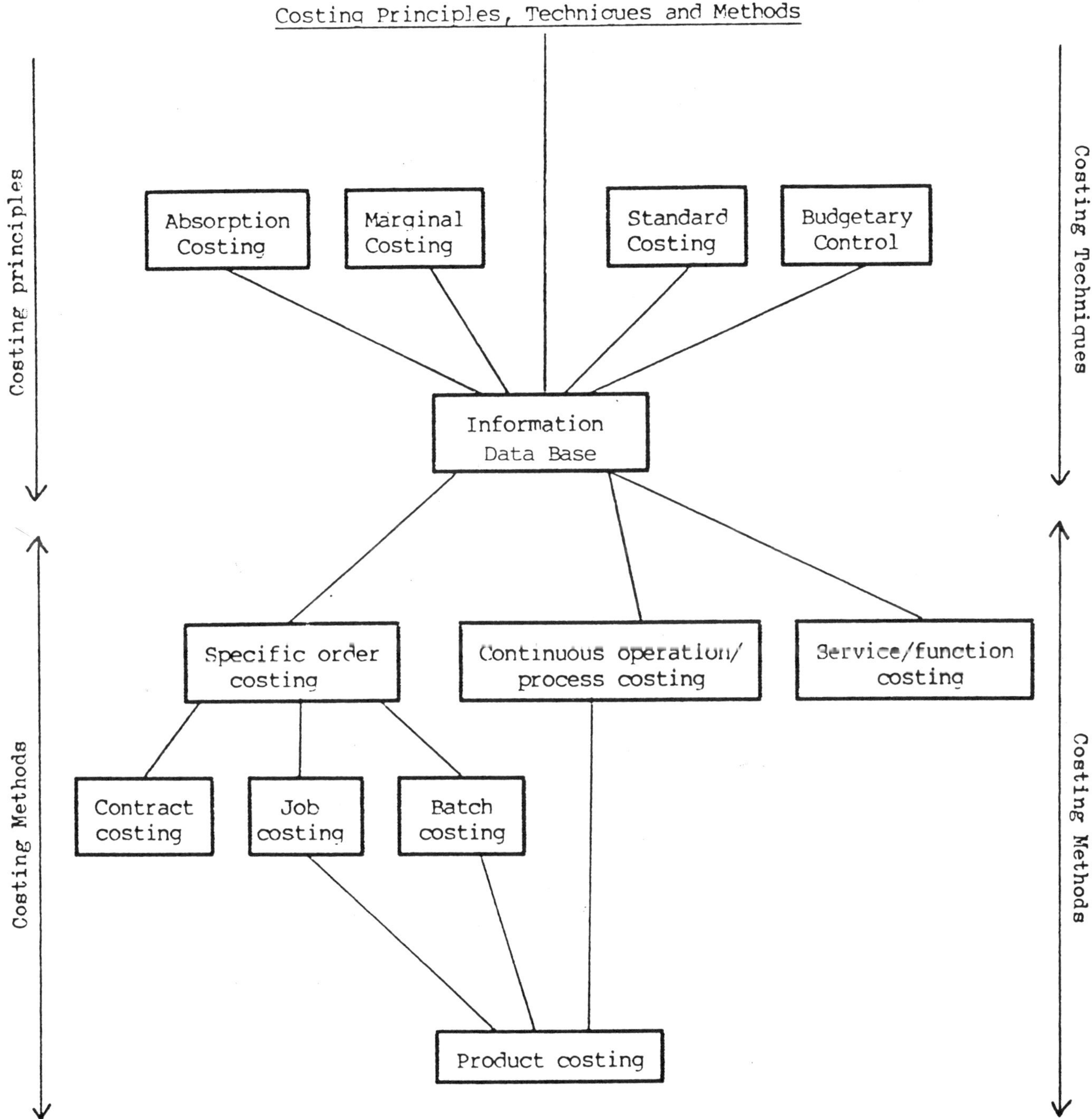

Figure 2

Accounting: Functional Relationship

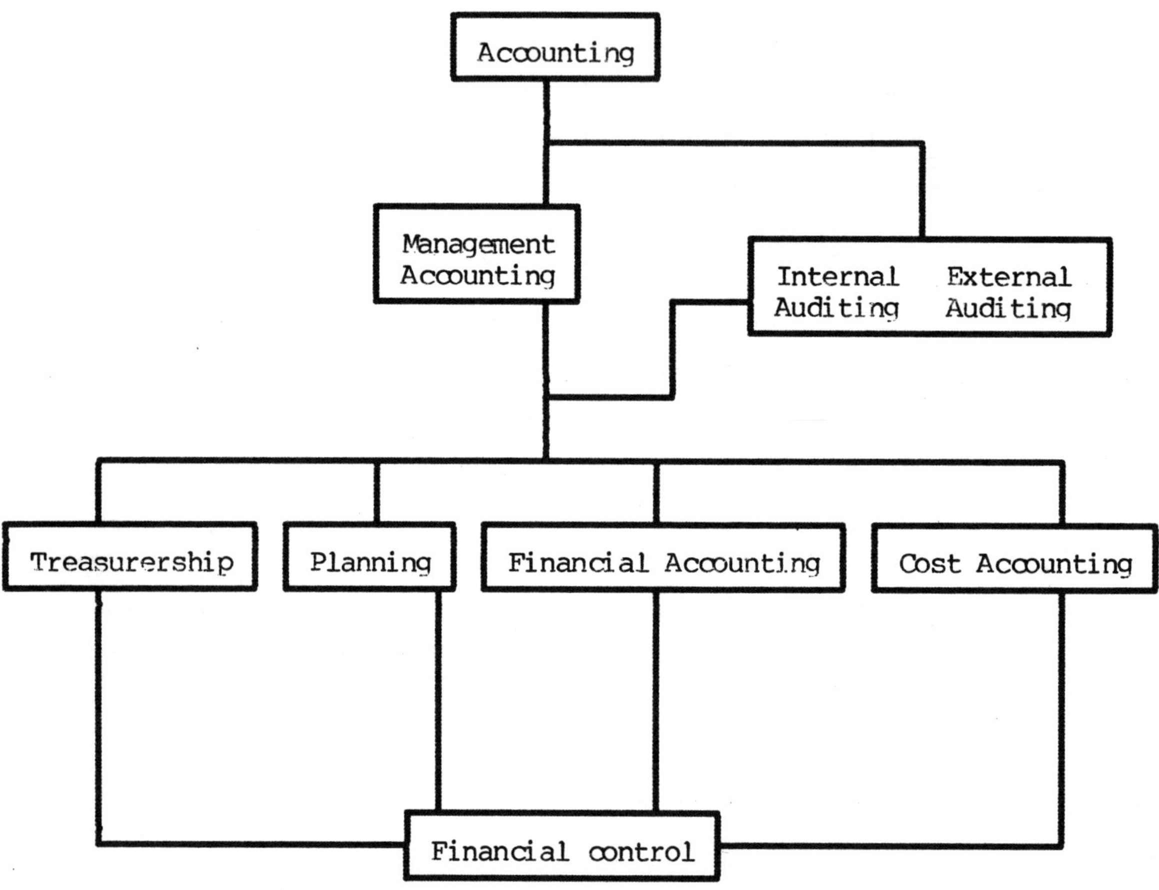

SELECTED REFERENCES AND BIBLIOGRAPHY

Accounting Standards Committee (UK) (1986) Accounting for the Effects of Changing Prices: A Handbook.

Accounting Standards Committee (UK) (1990) Cash Flow Statements - Exposure Draft 54.

Accounting Standards Steering Committee (1975) The Corporate Report.

American Institute of Certified Public Accountants (1981) Accounting by Agricultural Producers and Agricultural Cooperatives - Issues Paper.

Ansoff, H. Igor (1966) Corporate Strategy; Penguin.

Anthony, Robert N and Dearden, John (1976) Management Control Systems; Irwin.

Bailey, M.J. "Formal Criteria for Investment Decisions"; Journal of Political Economy, 1959.

Baxter, W.T. (1984) Inflation Accounting; Philip Allan.

Bierman Jr., Harold and Smidt Seymour (1984) The Capital Budgeting Decision; MacMillan Publishing Company.

Boersema, John M. (1977) Capital Budgeting Practices including the Impact of Inflation - A Research Study; The Canadian Institute of Chartered Accountants.

Briston, R.J. (1982) Introduction to Accountancy and Finance; London Institute of Cost & Management Accountants.

Brown, J.L. and Howard L.R. (1982) Managerial Accounting and Finance; McDonald and Evans Ltd.

Brown, C.I., Mallet, D.J. and Taylor, M.G. (1983) Banks, An Accounting and Auditing Guide; The Institute of Chartered Accountants in England and Wales.

Canadian Institute of Chartered Accountants (1986) Accounting and Financial Reporting by Agricultural Producers - A Research Study.

Carsberg, B. and Hope, A. (1976) Business Investment Decisions under Inflation; Institute of Chartered Accountants in England and Wales.

Commonwealth Secretariat — A Manual on Project Planning for Small Economies; May 1982.

Dasgupta, P, Marglin, S. and Sen, A. — (1972) Guidelines for Project Evaluation; United Nations.

Davis, Edward and Pointon, John — (1984) Finance and the Firm; Oxford University Press.

Dewhurst, R.F.J. — Business Cost-Benefit Analysis; McGraw-Hill.

Duvigneau, J. Christian Prasad, Ranga N. — (1984) Guidelines for Calculating Financial and and Economic Rates of Return for DFC Projects; World Bank.

Egginton, D.A. — (1977) Accounting for the Banker; Longman.

Gittinger, J. Price — (1982) Economic Analysis of Agricultural Projects; The John Hopkins University Press.

Goldschmidt, Yaaqov, Shashua, Leon and Hillman, Jimmye S. — (1986) The Impact of Inflation of Financial Activity in Business, with Applications to the U.S. Farming Sector; Rowman and Littlefield.

Granger, Charles H. — "The Hierarchy of Objectives"; Harvard Business Review, May-June 1964.

H.M. Treasury Advisory Group (UK) — (1986) Accounting for Economic Costs and Changing Price - Volumes I and II.

Hague, D.C. — (1971) Pricing Decisions; George Allen & Unwin.

Haley, Charles W. and Schall, Lawrence — (1979) The Theory of Financial Decisions; McGraw-Hill.

Hardie, J.D.M. — (1976) A Guide to Basic Agricultural Project Appraisal in Developing Countries; Aberdeen (UK).

Harvey, C. — (1983) Analysis of Project Finance in Developing Countries; Heinemann.

Hicks, J.R. — (1946) Value and Capital; Oxford Clarendon Press.

Hirshliefer, J. — "Theory of Optimal Investment Decisions", Journal of Political Economy, 1958.

International Accounting Standards Committee — (1990) Disclosures in the Financial Statements of Banks and Similar Financial Institutions (International Accounting Standard 30)

International Accounting Standards Committee — (1990) International Accounting Standards (issued up to 1 January 1990).

International Federation of Accountants — (1989) Proposed International Statement on Auditing - - The Audits of International Commercial Banks; Exposure Draft dated 17 April 1989.

Lee, Tom — (1984) Cash Flow Accounting; Van Nostrand Reinhold (UK) Co. Ltd.

Llewellyn, D.T. — (1986) The Regulation and Supervision of Financial Institutions; The Institute of Bankers (Gilbart Lectures on Banking).

Lumby, Stephen — (1984) Investment Appraisal; Van Nostrand Reinhold (UK) Co. Ltd.

Merret, A.J. and Sykes, A. — (1973) The Finance and Analysis of Capital Projects; Longman.

Mishan, E.J. — (1982) Cost-Benefit Analysis; George Allen & Unwin.

Mishkin, Frederic S. — (1986) The Economics of Money, Banking, and Financial Markets; Little Brown and Company.

Modigliani, F. and Miller, M.H. — "The Cost of Capital, Corporation Finance and the Theory of Investment"; American Economic Review, June 1958.

Mould, Maurice C. — (1989) Financial Information for Management of a Development Finance Institution - Some Guidelines; World Bank Technical Paper No. 63, Industry and Finance Series.

Orgler, Yair E. and Wolkowitz, Benjamin — Bank Capital; Van Nostrand Reinhold Company.

Prest, A.R. and Turvey, R. — "Cost-Benefit Analysis, A Survey"; Economic Journal, 1965.

Reid, Walter and Myddelton, D.R. — (1974) The Meaning of Company Accounts; Gower Press.

Report of the Committee to Review the Functioning of Financial Institutions; (June 1980), London HMSO.

Report of the Inflation Accounting Committee; (Sept. 1975), London HMSO.

Revell, J.R.S. — (1980) Costs and Margins in Banking; OECD Paris.

Samuels, J., Rickwood, C. and Piper, A. — (1981) Advanced Financial Accounting; McGraw-Hill.

Sizer, John — An Insight into Management Accounting; Penguin.

Sloane, Leslie — (1985) Accounting in British Banking; The Institute of Cost and Management Accountants.

Solomon, E. — "The Arithmetic of Capital Budgeting Decisions"; Journal of Business, 1956.

Turvey, R. — "On Divergencies between Social Cost and Private Cost"; Economica, 1963.

Weston, J. Fred and Brigham, Eugene F. — (1989) Managerial Finance; Holt Reinhart and Winston.

Wood Jr., Oliver G. and Porter, Robert J. — Analysis of Bank Financial Statements; Van Nostrand Reinhold Company.

World Bank — (1989) Report of the Task Force on Financial Sector Operations.

World Bank — Draft Operational Directive on Financial Sector Operations, dated 17 April 1990 (unpublished).